Ten Lessons on Risks

Xingwei Zhen · Jan-Erik Vinnem

Ten Lessons on Risks

Principles, Methods and Applications in the Petroleum Sector

Springer

Xingwei Zhen
School of Naval Architecture and Ocean Engineering
Dalian University of Technology
Dalian, China

Jan-Erik Vinnem
Department of Marine Technology
Norwegian University of Science and Technology
Trondheim, Norway

ISBN 978-3-031-99063-2 ISBN 978-3-031-99064-9 (eBook)
https://doi.org/10.1007/978-3-031-99064-9

This Springer imprint is published by the registered company Springer Nature Switzerland AG
The registered company address is: Gewerbestrasse 11, 6330 Cham, Switzerland

Those who cannot change their minds cannot change anything.

—George Bernard Shaw

For a safer future

Preface

In today's era, when science and technology flourish, risk assessment has emerged as a predominant tool in the management of Health, Environment, and Safety (HES) across various industries worldwide. In particular, from an exploratory start in Norway, Quantitative Risk Assessment (QRA) has become a cornerstone of modern safety protocols in the petroleum sector. Nonetheless, this tool is not easy to understand and master for beginners and even professionals due to the non-uniform terminologies, confusing concepts, various analysis methods, as well as the cross and fusion of multi-disciplinary knowledge. What is particularly worrying is that the fundamental purpose of risk assessment is frequently forgotten, and what is worse, risk assessment is employed to justify deviations from regulatory standards or established industry practices. Thus, risk assessment may degenerate into "safety on paper". The persistent occurrence of major accidents has substantiated such worries.

The author, Jan-Erik Vinnem, has accomplished a monumental work, *Offshore Risk Assessment: Principles, Modelling and Applications of QRA Studies* (Vinnem and Røed 2020), which is the first textbook to address QRA as specifically applied to offshore installations and operations. First published in 1999, this pioneering work has undergone successive revisions in 2007, 2014 and 2020. Notably, the second edition of this book, designated as a National Key Publication under the "Twelfth Five-Year Plan", was translated into Chinese and officially introduced to China in 2012. Nevertheless, no subsequent updates to the Chinese edition have been released thereafter, resulting in the omission of many important developments in the field of offshore risk assessment since 2007. Meanwhile, as university professors, the authors notice that people, while accustomed to fragmented reading in the new media age, often find it challenging to engage with this weighty two-volume work from cover to cover.

In response to the aforementioned challenges, this book has been conceived as a systematic exploration of solutions. With great ambition, the authors would like to provide a comprehensive overview of risk management in accordance with ten lessons, with a focus on its principles, methods, and applications in the petroleum sector. It is worth noting that the Chinese version of this book is currently undergoing expedited preparation, driven by the aspiration to enhance risk awareness in the Chinese petroleum industry.

Considering the authors' constraints on time and energy, some errors or omissions in this book are inevitable. The readers are welcome to raise any questions or feedback pertaining to this book. The authors hold regular monthly meetings, and are happy to address these inquiries timely during their scheduled meetings.

Organization of the Book

This book presents a comprehensive introduction to risk management, concentrating on its principles, main methods, and applications in the petroleum sector. The emphasis is placed on the major accident risk of socio-technical systems. The book is structured into the following ten chapters.

Chapter 1 provides readers with the necessary knowledge on how to understand risks. In Chap. 2, readers will have a further understanding of what risk management is and the different steps that make out the process. This is followed by Chap. 3, where the importance of learning from past accidents is demonstrated and the methods for accident investigations are provided. Chapter 4 presents the main techniques for different phases of risk analysis. Chapters 5 and 6 discuss the principles of risk-based design as well as risk-based emergency response planning. In Chaps. 7 and 8, the establishment of barriers and indicators for major accident risk is discussed. The decision framework for risk management is elaborated in Chap. 9. Chapter 10 discusses the most important topic of this book pertaining to ethics in risk assessment. At the end of each lesson, a series of questions is provided for readers to prompt deeper thinking and enable self-evaluation of lesson comprehension.

Who Is Our Audience?

This book will be of interest to a wide audience, including researchers, educators, and college students, experts in the industry, enterprise managers, consultants and regulators in the field of risk management. The book aspires to serve as a practical guide to effective risk management.

The book also appeals to the general public with an interest in risks. In a world where uncertainty is the only certainty, mastering essential risk knowledge and enhancing risk awareness are crucial for transforming uncertainty from a challenge into an opportunity, thus empowering informed decision-making.

It is our sincere hope that readers will enjoy a fresh and engaging reading experience with this book.

Dalian, China
Trondheim, Norway
May 2025

Xingwei Zhen
Jan-Erik Vinnem

Reference

Vinnem JE, Røed S (2020) Offshore risk assessment: principles, modelling, and applications, 4th edn. Springer, London

Acknowledgments Sincere thanks of the authors are expressed to the following researchers and friends who have provided valuable assistance in the preparation of this book: Bjørnar Heide, risk management specialist at Havtil; Yue Han, Ph.D. student at Dalian University of Technology (DUT).

The authors once initiated a pioneering project in China to develop the Intelligent Operational Safety Management System for offshore installations. Our sincere thanks are extended to the following collaborators, whose contributions have been beneficial to certain sections of this book: Prof. Ali Mosleh, University of California, Los Angeles (UCLA), National Academy of Engineering; Prof. Jørn Vatn, Norwegian University of Science and Technology (NTNU), Norwegian Academy of Technological Sciences. Havtil generously supported the use of its Barrier Memorandum, and DUT contributed to the publication of this book through a Textbook Development Fund (JC2024004).

Xingwei Zhen wishes to express his deepest gratitude to his tutor, colleague, and friend, Prof. Jan-Erik Vinnem, who always pours out his vast knowledge and valuable experience without holding back, as well as embraces every challenge with enthusiasm and open arms. Here, Xingwei Zhen would like to express his feeling in accordance with a statement from the preface of Yu-Tang Lin's book *The Importance of Living (1937)*: "The merits of this book, if any, are largely due to the helpful suggestions of my collaborators, while for the inaccuracies, deficiencies and immaturities of judgment, I alone am responsible."

Sincere thanks also go to our families, whose quiet love and unwavering support have provided the foundation for this book.

Competing Interests The authors have no competing interests to declare that are relevant to the content of this manuscript.

About This Book

This book offers a comprehensive guide to risk management in the petroleum sector through a structured curriculum of ten lessons, designed to progressively equip readers with both foundational knowledge and advanced skills. Beginning with the basics of risk understanding, it advances to in-depth exploration of the risk management process, accident investigations, and risk modelling techniques. Subsequently, the book delves into risk-based design, risk-based emergency response planning, as well as the establishment of barriers and indicators for major accident risk. Further, the book presents and discusses the decision-making framework for risk management. The book culminates in the most important topic on ethics in risk assessment, ensuring that the technical analyses align with societal values, justice, and long-term safety. Each lesson concludes with thought-provoking questions to foster critical thinking and self-assessment, reinforcing practical application of relevant concepts, principles, and methods. In particular, the most recent NORSOK Z-013 standard is followed in this book with respect to the processes of risk assessment and emergency preparedness assessment.

This book serves as a practical reference guide for researchers, educators and college students in risk management, empowering them to navigate complex risk issues with confidence and ethics. It will also be of interest to experts in the industry, enterprise managers, consultants and regulators.

Contents

About the Authors

Xingwei Zhen is an Associate Professor of ocean engineering at Dalian University of Technology. He has more than 15 years of experience in academic research, consulting, teaching as well and project management. He has authored more than 100 scholarly works, including peer-reviewed publications, granted patents, and software copyrights. He has been working together with Prof. Jan-Erik Vinnem since the year 2018 in the fields of operational risk modelling, online risk analysis, barrier management, major accident risk indicators and accident investigations. In addition, Xingwei Zhen has broad scientific interests. He invented an artificial seabed unit to extend the application of shallow-water-rated, cost-effective subsea production systems in ultra-deep water. He has also independently designed and constructed a large-scale internal wave experimental flume to explore and demystify the unique hydrodynamic phenomena of internal waves, as well as the mechanisms by which they pose threats to deepwater engineering facilities. Despite these achievements, his true passion lies in applied risk studies that he views as meaningful and beneficial to society.

Jan-Erik Vinnem is a world-renowned scholar of operational offshore risk and a professor at Norwegian University of Science and Technology, as well as a member of the Norwegian Academy of Technological Sciences. He is the founder of Safetec Nordic AS and Preventor AS. He is also the founding Chairman and an honorary member of ESAR Norway and the chairman of ESREL conferences in 1998, 2007, and 2018. He is the project leader for the development of OLF Guideline 064, ISO 17776 and NORSOK Z-013 standards. His main research topics during the last 10 years have been management of barriers and monitoring of major accident risk, with extensive projects for Havtil and petroleum companies. He is the author of seven books on risk analysis and has published more than 500 scientific papers and featured articles in international journals, conferences, and magazines. He is a recognized lecturer in risk management and risk analysis, and has conducted training courses for corporate clients in Norway as well as for international clients and research institutions across Denmark, Portugal, Brazil, Nigeria, India, Russia and China.

Abbreviations

AAIB	Air Accidents Investigation Branch
AAS	Autonomous Assistant System
AI	Artificial Intelligence
AIR	Average Individual Risk
ALARP	As Low As Reasonably Practicable
ALS	Accidental Limit States
AMF	Automatic Mode Function
ANP	National Agency for Petroleum, Natural Gas and Biofuels (Brazil)
APF	Average Percentage of Failures
BBD	Barrier Block Diagram
BDV	Blowdown Valve
BFETS	Blast and Fire Engineering for Topside Structures
BN	Bayesian Network
BOP	Blowout Preventer
BORA	Barrier and Operational Risk Analysis
BSEE	Bureau of Safety and Environmental Enforcement
CAA Norway	Civil Aviation Authority of Norway
CBA	Cost-Benefit Analysis
CCR	Central Control Room
CCTV	Closed-Circuit Television
CDSM	Cidade de São Mateus
CED	Cause-and-Effect Diagram
CFD	Computational Fluid Dynamics
CM	Corrective Maintenance
CNOOC	China National Offshore Oil Corporation
COPC	ConocoPhillips China
CPT	Conditional Probability Table
CSB	Chemical Safety Board
DAG	Directed Acyclic Graph
DBN	Dynamic Bayesian Network
DFU	Defined Hazards and Accident Conditions

DHSV	Downhole Safety Valve
DiAE	Dimensioning Accidental Event
DiAL	Dimensioning Accidental Load
DMA	Danish Maritime Authority
DOE	Department of Energy
DSB	Directorate for Civil Protection and Emergency Planning
DSHA	Defined Situation of Hazard and Accident
DWH	Deepwater Horizon
ECF	Event and Causal Factor
EDS	Emergency Disconnect System
EER	Evacuation, Escape and Rescue
EIA	Environmental Impact Assessment
EPA	Emergency Preparedness Assessment
ERRV	Emergency Response and Rescue Vessels
ERS	Emergency Response Strategy
ESD	Event Sequence Diagram
ESDV	Emergency Shutdown Valve
ETA	Event Tree Analysis
EU	European Union
F-AHP	Fuzzy Analytic Hierarchy Process
FAR	Fatal Accident Rate
FBD	Functional Block Diagram
FEED	Front-End Engineering and Design
FES	Fire and Explosion Strategy
FI	Frequency of Impairment
FLACS	Flame Accelerator Software
FMEA	Failure Modes and Effects Analysis
FMECA	Failure Modes, Effects, and Criticality Analysis
FPSO	Floating Production, Storage and Offloading
FRC	Fast Rescue Craft
FTA	Fault Tree Analysis
GoM	Gulf of Mexico
GR	Group Risk
Havtil	Norwegian Ocean Industry Authority
HAZAN	Hazard Analysis
HAZID	Hazard Identification
HAZOP	Hazard and Operability
HEP	Human Error Probability
HES	Health, Environment, and Safety
HMI	Human Machine Interface
HOF	Human and Organizational Factor
HPES	Human Performance Enhancement System
HSE	Health and Safety Executive
IAEA	International Atomic Energy Agency
IEC	International Electrotechnical Commission

IR	Individual Risk
ISO	International Organization for Standardization
JMC	Joint Management Committee
LAHH	Level Alarm High High
LALL	Level Alarm Low Low
LEL	Lower Explosive Limit
LMRP	Lower Marine Riser Package
LNG	Liquefied Natural Gas
LTI	Lost-Time Injuries
MAH	Major Accident Hazard
MCS	Minimal Cut Set
MGS	Mud Gas Separator
MOC	Management of Change
MODU	Mobile Offshore Drilling Unit
MSF	Main Safety Function
MTO	Man-Technology-Organization
NATO	North Atlantic Treaty Organization
NCS	Norwegian Continental Shelf
NOPSEMA	National Offshore Petroleum Safety and Environmental Management Authority
NPT	Negative-Pressure Test
NSIA	Norwegian Safety Investigation Authority
NSOPB	Nova Scotia Offshore Petroleum Board
NTSB	National Transportation Safety Board
OCS	Outer Continental Shelf
ODP	Overall Development Program
OMS	Organizational Management System
OPF	Overall Percentage of Failures
ORM	Online Risk Models
ORMDS	Online Risk Monitoring and Decision Support
OTS	Operational Condition Safety
PAHH	Pressure Alarm High High
PDO	Plans for the Development and Operation
PEAR	People, Environment, Asset and Reputation
PFD	Probability of Failure on Demand
PHA	Preliminary Hazard Analysis
PID	Piping and Instrumentation Diagrams
PL	Penglai
PLL	Potential Loss of Life
PM	Preventive Maintenance
PR	Performance Requirements
PRA	Preliminary Risk Analysis
PS	Performance Standards
PSC	Production Share Contract
PSD	Process Shutdown

PSV	Pressure Safety Valve
PTW	Permit-to-Work
QRA	Quantitative Risk Analysis
RAC	Risk Acceptance Criteria
RBDM	Risk-Based Decision-Making
RID	Risk Influencing Diagrams
RIDM	Risk-Informed Decision-Making
RIF	Risk Influencing Factor
Risk_OMT	Risk Modelling: Integration of Organizational, Human and Technical Factors
RNNP	Trends in Risk Level in the Norwegian Petroleum Activity
ROV	Remotely Operated Vehicles
RPN	Risk Priority Number
RTC	Risk Tolerance Criteria
SAR	Search and Rescue
SCR	Safety Case Regulations
SFAIRP	So Far As Is Reasonably Practicable
SIF	Safety Instrumented Function
SIL	Safety Integrity Level
SIS	Safety Instrumented System
SOA	State Oceanic Administration
SRA	Society for Risk Analysis
SSIV	Subsea Isolation Valve
STEP	Sequential Timed Events Plotting
TAHH	Temperature Alarm High High
TALL	Temperature Alarm Low Low
TR	Temporary Refuge
TSB	Transportation Safety Board
UFM	Uncontrolled Forward Movement
UK	United Kingdom
US	United States

First Lesson: Risk Understanding

1

The words of risk analysis have been, and continue to be a problem. Many of you here remember that when our Society for Risk Analysis was band new, one of the first things it did was to establish a committee to define the word "risk". This committee labored for four years and then gave up, saying in its final report, that maybe it is better not to define risk. Let each author define it in his own way, only please each should explain clearly what way that is.

—*Kaplan* (1997)

Abstract

The concept of "risk" lies at the heart of decision-making processes across industries, yet its interpretation and application often vary depending on the context, objectives, and stakeholder perspectives. This lesson seeks to demystify the multifaceted nature of risk by exploring its definition, description and measurement. Building on this, the purpose of risk assessment is highlighted as it is easily forgotten. The lesson further explores the crucial frameworks of the Norwegian, UK (United Kingdom) and EU (European Union) regulations and relevant standards, which provide structured methodologies and compliance requirements to ensure systematic and consistent risk management practices. After that, uncertainties and sensitivities, which are tightly linked but distinct concepts, are discussed. This discussion extends to black swan risks, which has attracted significant public attention. The authors argue that the concept of black swan risk should be restricted to the realm of "unknown unknowns". Lastly, a key focus is placed on Human and Organizational Factors (HOFs) and their significant influence on risk. By synthesizing these elements, the lesson aims to equip readers with a holistic understanding of risk.

X. Zhen and J.-E. Vinnem, *Ten Lessons on Risks*,
https://doi.org/10.1007/978-3-031-99064-9_1

1.1 What Is Risk?

This section aims to clarify the concept of risk through four interconnected dimensions: risk definition, risk description, the bow-tie model for visualizing risk, as well as risk measurement. In this book, beyond the conventional "operational definition" of risk (i.e., the combination of likelihood and magnitude of consequences for a specific accident scenario), risk is further described by addressing five critical questions:

(i) What can go wrong?
(ii) How likely is that to happen?
(iii) What are the consequences?
(iv) What are the uncertainties?
(v) How strong is the strength of knowledge?

1.1.1 The Definition of Risk

As noted by the prominent risk researcher Kaplan (1997), a universally agreed-upon definition of risk is still regrettably absent, though significant time and efforts have been devoted by a number of scholars in this area. This challenge can be attributed to two primary reasons (Rausand and Haugen 2020. On the one hand, the concept of risk is studied across a multitude of disciplines within the fields of social, natural and technical science, each contributing distinct perspectives, terminologies and foundational approaches. On the other hand, the term "risk" has come into everyday language, often resulting in ambiguity and a lack of precise meaning.

In response to this challenge, the Society for Risk Analysis (SRA) compiled a lexicon of terminologies for risk analysis (SRA 2018). The SRA seeks to provide a general and consistent understanding of essential concepts within this field. Rather than offering a universal definition of risk, the glossary presents a series of qualitative definitions, some of which are outlined below:

(1) Risk is the possibility of an unfortunate occurrence.
(2) Risk is the potential for the realization of unwanted, negative consequences of an event.
(3) Risk is the consequences of the activity and associated uncertainties.
(4) Risk is exposure to a proposition that is uncertain.
(5) Risk is the deviation from a reference value and associated uncertainties.

Additionally, the International Standards Organization (ISO) also provides a widely recognized definition of risk, which refers to the effect of uncertainty on objectives (ISO 2009).

It is noteworthy that the concept of risk within the Norwegian offshore petroleum sector has undergone an evolutionary process. Historically, the risk concept has been dominated by a probabilistic approach over the past few decades (Vinnem and Røed 2020; Røyksund and Engen 2020). This probabilistic approach to risk often relies exclusively on accident statistics. Nonetheless, effective risk management necessitates a number of assumptions, experience and accumulated knowledge, which have frequently been neglected by the probabilistic approach. For this reason, the challenge lies in how to determine whether the assumptions and expertise are robust enough to encompass all relevant factors. Consequently, this probabilistic approach based on statistics in defining risk has been criticized for its narrow scope, as it inadequately addresses the dimensions of uncertainty and the strength of knowledge, both of which are crucial during the decision-making process. In response to all these critiques, the Norwegian Ocean Industry Authority, referred to as Havtil in Norwegian, introduced an uncertainty-based risk perspective in the year 2015. This perspective defines risk as "the consequences of the activities, with associated uncertainties" (Havtil 2015).

The definitions offered by both ISO and Havtil align with a specific instance of definition (3) presented in the SRA glossary. Note that both of these definitions in principle allow positive consequences, not restricted to negative consequences, but this is usually disregarded. Some views would like to link the risk concept with the negative consequences. Nonetheless, it is wise not to make this linkage, as whether the consequences are negative or positive is a relative concept to different stakeholders. In this book, the risk definition of Havtil is supported.

In quantitative risk analysis, the "operational definition" of risk (R) is commonly applied and expressed as the expected consequences, i.e., the combination of likelihood (P) and magnitude of consequences (C) for a specific accident scenario i:

$$R = \sum_{i} (P_i \times C_i) \tag{1.1}$$

The risk metric discussed herein is a statistical representation, indicating that the numerical value may be unobserved. Consequently, it is necessary to establish an average dataset over an extended period, especially in the context of rare accidents. Furthermore, it is essential to acknowledge that this statistical expression can merely provide limited information for describing the risk picture. Additional information can be derived from studies, such as risk contributions, effects of changes, and risk improvements.

The risk metric denoted in Eq. (1.1) is commonly termed "statistical risk" or "technological risk". Some scholars also categorize this metric as "real risk" or "objective risk". However, from the perspective of authors, the terms "real" and "objective" may result in a misleading interpretation of risk. Risk is subjective to some extent due to its reflection on the interpretations and simplifications of the analysts. Therefore, it is misleading to suggest that certain risk metrics are inherently more "real" or "objective" than others.

Notably, this book focuses on risks associated with technical or socio-technical systems, specifically concerning sudden future events that could result in severe outcomes, i.e., potential major accidents.

1.1.2 The Description of Risk: Five Critical Questions

Risk is inherently associated with future activities. To further understand the definition of risk, five fundamental questions are proposed and explained (Kaplan 1997; Kaplan and Garrick 1981; Aven 2008; SRA 2018; Rausand and Haugen 2020; Vinnem and Røed 2020) to systematically describe the risk.

Q1: What can go wrong?

The question can be answered as the potential accident scenarios or accident sequences that may jeopardize the defined values. An accident scenario represents a process characterized by a sequence of events, commencing with a triggering event and culminating in a specified outcome. The defined values primarily encompass personnel, environment and economic assets within the petroleum engineering domain. In addressing this question, hazards, which represent potential sources of harm, are identified.

Q2: How likely is that to happen?

The response to this question involves determining the frequencies or probabilities associated with the accident scenarios, which are identified in the first question.

Q3: What are the consequences?

This question seeks to ascertain the final effects of the activity on the defined values as set forth in the first question, with a particular emphasis on the negative or undesirable consequences.

Q4: What are the uncertainties?

The answer to this question involves recognizing the imperfect or incomplete information and knowledge available for the analysis of each accident scenario. The various emerging definitions of risk underscore the importance of uncertainties. It is important to understand why both uncertainty and probability are needed. In other words, uncertainty cannot be directly replaced by probability. This is because the probability serves merely as one of the tools to express the uncertainty concerning the accident scenario and its consequences. In particular, it is not a perfect tool. As has been stated by the Health and Safety Executive (HSE[1] 2001) in the UK, uncertainty represents a state of knowledge in which, despite the identified factors influencing the issue, the

[1] The abbreviation 'HSE' is in this book used for the UK authority, whereas the term 'HES' (see Sect. 9.1) is used for the topic 'Health, Environment and Safety'.

likelihood of adverse effects or the effects themselves cannot be precisely described. Thus, certain uncertainties may not be adequately reflected by probability. Further discussions on this topic can be referred to Aven (2008).

Q5: How strong is the strength of knowledge?

The determination of frequencies or probabilities for the identified accident scenarios needs to be grounded in the background knowledge, which correlates to the justification of assumptions and limitations, the availability of reliable data and information, comprehension of relevant phenomena, as well as experience and testing. The important thing to realize is that the identical probability values derived by different analysts may stem from different background knowledge. This indicates that the risk metrics alone cannot provide sufficient decision support. Thus, it is essential to differentiate between "weak" and "strong" background knowledge, which reflects the robustness of the foundation for decision-making.

On the basis of the five questions previously outlined, the risk (R) associated with the conducted activity can be expressed in the following manner:

$$R = (S, P, C, U, K) \tag{1.2}$$

where

S represents the accident scenario,
P represents the likelihood of the accident scenario occurring,
C represents the consequences resulting from the accident scenario,
U represents the uncertainty regarding the estimated consequences,
K represents the strength of knowledge.

Notably, in some cases, risk is described as a synthesis of the likelihood of an accident occurring, the magnitude of its potential consequences, and the aversion linked to those consequences. "Risk aversion" describes the analysts' attitude toward risk, indicating that negative consequences are disliked so strongly that greater weights are assigned to these consequences in comparison to a statistical mean value approach. Thus, risk aversion is factored into the risk calculation process. Nonetheless, this interpretation of risk is not endorsed by the authors, though the importance of risk aversion in evaluating the risk analysis results is acknowledged. Risk aversion represents a complex phenomenon, and may misleadingly suggest that the complex process can be sufficiently represented by a singular risk aversion factor. The process should include evaluation of uncertainties in the analysis results as well as a consideration of ethical values and political and social considerations relating to the potential consequences. For a more comprehensive understanding of risk aversion, one may refer to the work of Aven and Vinnem (2007).

Now, on the basis of the five questions presented, consider how one might provide a description of the risk associated with stepping onto Kjeragbolten (as

Fig. 1.1 The Kjeragbolten in proximity to Stavanger, Norway (Photograph by the author, Xing-Wei ZHEN)

illustrated in Fig. 1.1), a famous rock suspended 1084 m over Lysefjorden between two cliffs.

1.1.3 Bow-Tie Model for Visualizing Risk

The bow-tie model is named after its shape, which resembles a bow-tie. It originated from the aviation industry and was developed to help personnel visualize the complex interactions among various elements that could result in undesirable events. Nowadays, the bow-tie model has been widely extended and applied across various industries.

Figure 1.2 presents a typical bow-tie model, which could be used to clearly visualize the risk concept in an understandable way. The central "knot" of the bow-tie model represents the undesirable (Top) event, while the left-most and right-most sides represent hazards and consequences, respectively. The bow-tie model indicates that hazards could evolve into the undesirable event, which in turn may result in further consequences. In order to minimize the probability of hazards developing or to mitigate the potential harm from an undesirable event, barriers are put in place on both sides of the model. It is worth noting that a barrier is a fundamental concept in risk modelling, denoting measures specifically designed to provide protection in hazardous and accidental situations. The Seventh Lesson of this book presents a comprehensive introduction to the concept of "barriers".

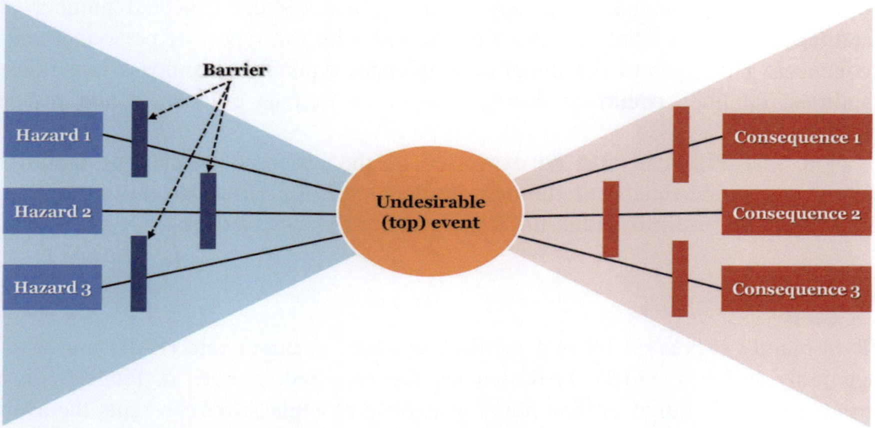

Fig. 1.2 A typical bow-tie model

The bow-tie model has become popular as a systematic approach for assessing risk, as well as illustrating the relationship among hazards, barriers, consequences and other related factors.

1.1.4 Measuring Risk

This section explores how to measure risk quantitatively, focusing on personnel risk, environmental risk, and economic asset risk.

Measuring Risk to Personnel

There are a number of ways to represent personnel risk. The discussion will focus on the most prevalent metrics, encompassing the potential loss of life (PLL), individual risk (IR), the *F-N* curve, and the frequency of impairment (FI). It is important to note that these risk metrics are restricted to fatalities, either directly or indirectly.

(1) PLL

PLL represents the statistically expected number of fatalities within a specified timeframe (per year in usual), and can be calculated by the equation outlined below

$$\mathrm{PLL} = \sum_i \sum_j \left(f_{ij} \cdot c_{ij}\right) \tag{1.3}$$

where

f_{ij} represents the annual frequency, while c_{ij} denotes the expected number of fatalities for the accident scenario i associated with the type j of personnel consequences. The types of personnel consequences typically encompass immediate fatalities, fatalities occurring during escape, as well as evacuation and rescue fatalities.

Note that PLL can also be expressed as the observed number of fatalities within a specified timeframe through accident statistics. However, this expression is typically not utilized within the framework of risk assessment.

(2) IR

IR is mainly expressed by two metrics: the fatal accident rate (FAR) and average individual risk (AIR). FAR denotes the expected number of fatalities in a group per one hundred million hours of exposure, while AIR represents the average number of fatalities per exposed individual. FAR and AIR can be calculated by the equations outlined below.

$$\mathrm{FAR} = \frac{\mathrm{PLL} \times 10^8}{\text{Exposed hours}} = \frac{\mathrm{PLL} \times 10^8}{\mathrm{POB_{av}} \times 8760} \tag{1.4}$$

$$\mathrm{AIR} = \frac{\mathrm{PLL}}{\text{Exposed individuals}} = \frac{\mathrm{PLL}}{\mathrm{POB_{av}} \times \frac{8760}{H}} \tag{1.5}$$

where

$\mathrm{POB_{av}}$	The average number of individuals on board in one year.
H	The number of offshore hours in one year per individual, including on-duty and off-duty hours.
8760	The number of hours in one year.
8760/H	The number of individuals needed to occupy a single offshore position.

Notably, when calculating FAR, there are two different ways: either by treating "exposure hours" as equivalent to "on-duty" hours (i.e., working hours), or by summing both "on-duty" and "off-duty" hours as "exposure hours". As offshore shifts are 12 h on-duty and 12 h off-duty, these two ways of calculating FAR differ by a factor of 2. It is therefore essential that, when FAR values are discussed, the basis on which they have been calculated is stated. By contrast, the calculation of AIR does not have this complication.

It can be found that Eqs. (1.4) and (1.5) demonstrate a strong correlation between FAR and AIR. Once the shift plan is determined, it becomes possible to derive one metric from the other.

(3) *F-N* curve

In contrast to the individual risk, group risk (GR) is used to express the accidents' influence on society as a whole. The *F-N* curve is typically employed as a tool for representing GR, where *F* denotes the cumulative frequency F(c) of a consequence $C \geq c$ (i.e., frequency of exceedance), and *N* represents the magnitude of the consequence, often measured by the number of fatalities. Both axes of the curve are displayed on a logarithmic scale. An example of the *F-N* curve is illustrated in Fig. 1.3. It is important to note that the *F-N* curve is always falling or flat due to the inherent characteristic of the cumulative frequency.

(4) FI

FI serves as a measure of risk to personnel indirectly. The determination of impairment frequencies can be derived from modelling physical responses to accidental loads. Thus, the calculation of accident consequences in terms of fatalities can be avoided. The approach to calculating impairment frequencies is closely linked to "temporary refuge (TR)" or "main safety function (MSF)", which have been addressed in the regulatory frameworks of both the UK and Norway, as follows

- TR (The Secretary of State 2015):
  - Life support safety function,
  - Command safety function,
  - TR Egress safety function,
  - Evacuation safety function.
- MSF (Havtil 2020b):
  - Escalation of accident situations,
  - Capacity of main load-bearing structures,
  - Rooms of significance to combating accidents,
  - Safe areas,
  - Escape routes.

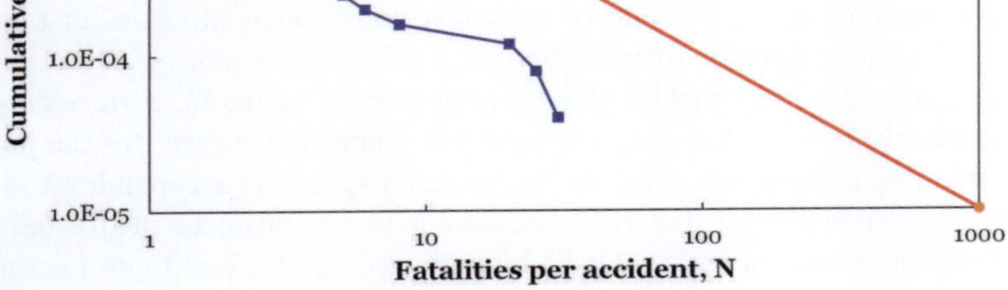

Fig. 1.3 Example of the *F-N* curve

The FI can be calculated by the equation outlined below.

$$\mathrm{FI} = \sum_{n} \left(f_n \times p_{\mathrm{imp},n,i} \right) \tag{1.6}$$

where

$p_{\mathrm{imp},n,i}$ denotes the probability of impairment associated with accident scenario n concerning the MSF i.

Measuring Risk to the Environment

The environmental risk primarily arises from the spills caused by blowouts and leaks in risers, pipelines and storage facilities. Though process leaks occur frequently, severe damage to the environment is not expected normally. The environmental risk is primarily expressed through the following metrics, and the consequence is typically measured in terms of restoration time.

- Expected spilled amount,
- The accumulated frequency of events with similar consequences for the environment.

Measuring Risk to Economic Assets

The economic risk is mainly expressed in the following ways. Nonetheless, the economic loss is commonly measured by means of monetary values.

- Expected structural and equipment damage,
- Expected duration of production delay,
- The accumulated frequency of events with similar consequences, either in the extent of structural and equipment damage or the duration of production delay.

1.2 Purpose of Risk Assessment

The purpose of risk assessment is to establish a sufficient and appropriate foundation for decision-making at the right time, encompassing the choices of solutions as well as measures for risk reduction. This fundamental purpose is the most explicit in the documentation requirements of the safety case, which stipulates that the risk reduction decisions must be justified by risk assessment results.

Nonetheless, this fundamental purpose is frequently forgotten, as the enterprises may mistakenly believe that risk assessment is conducted merely for the purpose of documenting that the risk level for the assessed system is acceptable. And what is worse, it is concerning that risk assessment is employed to justify deviation from regulatory standards or established industry practices. The Tenth Lesson will provide an in-depth discussion on the misuse of risk assessment.

What calls for special attention is that there are two distinct approaches to decision-making based on risk assessment: risk-based decision-making (RBDM) and risk-informed decision-making (RIDM). The differences between these two decision-making approaches are rooted in their focus. The RBDM approach emphasizes primarily the results of risk assessment, whereas the RIDM approach takes a more comprehensive view by incorporating the results of risk assessment alongside additional information, covering deterministic analysis and technical considerations. The Ninth Lesson will further elaborate on this topic.

1.3 Regulations and Standards

Several national authorities mandate the implementation of QRAs to ensure offshore safety. This section provides a concise overview of important regulations and standards within the offshore petroleum sector, including those established in Norway, the UK, the EU, and other regulatory regimes. In particular, the Norwegian regulatory text pertaining to risk assessment is presented as it stood at the end of 2023.

1.3.1 Norwegian Regulations

Havtil has issued four regulations as follows, with the Framework Regulations serving as the overarching regulations that outline the fundamental principles for the other three regulations.

- The Framework Regulations (Havtil 2019): These regulations address health, safety and the environment within the petroleum activities and certain onshore facilities.
- The Management Regulations (Havtil 2020a): These regulations address management and the duty to provide information in relation to the petroleum activities and certain onshore facilities.
- The Facilities Regulations (Havtil 2020b): These regulations pertain to the design and outfitting of facilities involved in the petroleum activities.
- The Activities Regulations (Havtil 2020c): These regulations outline the requirements for conducting petroleum activities.

The regulatory text pertaining to risk assessment in the Framework Regulations, Management Regulations, and Facilities Regulations is presented in the following. With respect to the Activities Regulations, there are no relevant requirements on risk assessment.

The Framework Regulations

Section 11 of the Framework Regulations outlines the risk reduction principles. In particular, the equivalent ALARP (As Low As Reasonably Practicable) evaluation in Norway is presented.

Harm or danger of harm to people, the environment or material assets shall be prevented or limited in accordance with the health, safety and environment legislation, including internal requirements and acceptance criteria that are of significance for complying with requirements in this legislation. In addition, the risk shall be further reduced to the extent possible.

In reducing the risk, the responsible party shall choose the technical, operational or organizational solutions that, according to an individual and overall evaluation of the potential harm and present and future use, offer the best results, provided the costs are not significantly disproportionate to the risk reduction achieved.

If there is insufficient knowledge concerning the effects that the use of technical, operational or organizational solutions can have on health, safety or the environment, solutions that will reduce this uncertainty shall be chosen.

Factors that could cause harm or disadvantage to people, the environment or material assets in the petroleum activities shall be replaced by factors that, in an overall assessment, have less potential for harm or disadvantage.

Assessments as mentioned in this section, shall be carried out during all phases of the petroleum activities.

This provision does not apply to the onshore facilities' management of the external environment.

The Management Regulations

Several sections of the Management Regulations are crucial for the analysis of barriers, the analysis of risk, and the establishment of risk acceptance criteria. Thereinto, the regulatory requirements for barriers are elaborated in the dedicated lesson on barriers for major accident risk, specifically the Seventh Lesson of this book, and will not be reiterated here.

Section 9:

Acceptance criteria for major accident risk and environmental risk

The operator and the party responsible for operating a mobile facility shall set acceptance criteria for major accident risk and for environmental risk associated with acute pollution.

Acceptance criteria shall be set for:

(a) the personnel on the offshore or onshore facility as a whole, and for personnel groups exposed to particular risk,
(b) loss of main safety functions for offshore petroleum activities,
(c) acute pollution from the offshore or onshore facility,
(d) damage to third party.

The acceptance criteria shall be used when assessing results from risk analyses, cf. Section 17. Cf. also Section 11 of the Framework Regulations.

Section 16:

General requirements for analyses

The responsible party shall ensure that analyses are carried out that provide the necessary basis for making decisions to safeguard health, safety and the environment. Recognized and suitable models, methods and data shall be used when conducting and updating the analyses.

The purpose of each risk analysis shall be clear, as well as the conditions, premises and limitations that form its basis.

The individual analysis shall be presented such that the target groups receive a balanced and comprehensive presentation of the analysis and the results.

The responsible party shall set criteria for carrying out new analyses and updating existing analyses as regards changes in conditions, assumptions, knowledge and definitions that, individually or collectively, influence the risk associated with the activities.

The operator or the party responsible for operating an offshore or onshore facility shall maintain a comprehensive overview of the analyses that have been carried out and are underway. Necessary consistency shall be ensured between analyses that complement or expand upon each other.

Section 17:

Risk analyses and emergency preparedness assessments

The responsible party shall carry out risk analyses that provide a balanced and most comprehensive possible picture of the risk associated with the activities. The analyses shall be appropriate as regards providing support for decisions related to the upcoming processes, operations or phases. Risk analyses shall be carried out to identify and assess what can contribute to, i.e., major accident risk and environmental risk associated with acute pollution, as well as ascertain the effects various processes, operations and modifications will have on major accident and environmental risk.

Necessary assessments shall be carried out of sensitivity and uncertainty.

The risk analyses shall:

(a) identify hazard and accident situations,
(b) identify initiating incidents and ascertain the causes of such incidents,
(c) analyze accident sequences and potential consequences, and,
(d) identify and analyze risk-reducing measures, cf. Section 11 of the Framework Regulations and Sections 4 and 5 of these regulations.

Risk analyses shall be carried out and form part of the basis for making decisions when e.g.:

(a) identifying the need for and function of necessary barriers, cf. Sections 4 and 5,
(b) identifying specific performance requirements of barrier functions and barrier elements, including which accident loads are to be used as a basis for designing and operating the installation/facility, systems and/or equipment, cf. Section 5,
(c) designing and positioning areas,
(d) classifying systems and equipment, cf. Section 46 of the Activities Regulations,
(e) demonstrating that the main safety functions are safeguarded,

(f) stipulating operational conditions and restrictions,
(g) selecting defined hazard and accident situations.

For larger discharges of oil or condensate, simulations of drift and dispersion shall be carried out.

Emergency preparedness analyses shall be carried out and be part of the basis for making decisions when e.g.

(a) defining hazard and accident situations,
(b) stipulating performance requirements for the emergency preparedness,
(c) selecting and dimensioning emergency preparedness measures.

The environmental risk and emergency preparedness analyses shall be updated in case of significant changes affecting the environmental risk or the emergency preparedness situation. In any case, updating needs shall be assessed every five years. The assessment shall be documented and made available to the Norwegian Environment Agency on request.

The Facilities Regulations

With respect to risk assessment, the Facilities Regulations focus on the maximum frequency of events that impair the main safety function, as outlined in the following. Notably, the main safety functions are more closely associated with design characteristics of the individual facility, as opposed to, for example, fatalities.

Section 7:

Main safety functions

The main safety functions shall be defined in a clear manner for each individual facility so that personnel safety is ensured and pollution is limited.

For manned facilities, the following main safety functions shall be maintained in the event of an accident situation:

(a) preventing escalation of accident situations so that personnel outside the immediate accident area are not injured,
(b) the capacity of main load-bearing structures until the facility has been evacuated,
(c) protecting rooms of significance to combatting accidents so that they remain operative until the facility has been evacuated,
(d) protecting the facility's safe areas so that they remain intact until the facility has been evacuated,
(e) at least one escape route from every area where personnel are found until evacuation to the facility's safe areas and rescue of personnel have been completed.

Section 11:

Loads, load effects and resistance

The design loads that will form the basis for design and operation of installations, systems and equipment, shall be determined. When determining design loads, the requirement to robust solutions, cf. Section 5, and the requirement to risk reduction, cf. the Framework Regulations Section 11, shall form the basis. The design loads shall ensure that installations, systems or equipment will be designed such that relevant accidental loads that can occur, do not result in unacceptable consequences, and shall, as a minimum, always withstand the dimensioning accidental load.

When determining design loads, the effects of fire water shall not be considered. This applies to both fire loads and explosive loads.

Installations, systems and equipment that are included as elements in the realization of main safety functions, cf. Section 7, shall as a minimum de designed such that dimensioning accidental loads or dimensioning environmental loads with an annual likelihood greater than or equal to $1x10^{-4}$, shall not result in loss of a main safety function.

When determining loads, the effects of seabed subsidence over, or in connection with the reservoir, shall be considered.

Functional and environmental loads shall be combined in the most unfavorable manner.

Facilities or parts of facilities shall be able to withstand the design loads and probable combinations of these loads at all times.

In addition, it can be noted that regulations pertaining to risk analyses for mobile offshore units have been established by the Norwegian Maritime Authority (NMA 2017). These risk analyses are required to be divided into several sub-analyses, which include concept analysis, design analysis, construction analysis, reliability/vulnerability analysis, and contingency analysis. The general requirements applicable to these risk analyses are also presented.

1.3.2 UK Regulations

Drawing upon the central recommendations outlined in the report of the public inquiry into the Piper Alpha disaster (Cullen 1990), the UK regulations on offshore safety were almost rewritten in the following

- OSCR 2005 (The Secretary of State 2005): The offshore installations (safety case) regulations.
- OSCR 2015 (The Secretary of State 2015): The offshore installations (offshore safety directive) (safety case, etc.) regulations.
- PFEER (The Secretary of State 1995a): The offshore installations (prevention of fire and explosion, emergency response) regulations.
- OMAR (The Secretary of State 1995b): The offshore installations and pipeline works (management and administration) regulations.
- ODCR (The Secretary of State 1996): The offshore installations and wells (design and construction etc.) regulations.

Particularly, the submission of safety cases constitutes a statutory requirement for all offshore installations currently operational or planned for deployment within the UK's territorial waters and its designated continental shelf jurisdictions.

A safety case represents a comprehensive document so as to demonstrate three fundamental elements: (a) identification of all hazards that could potentially result in a major accident; (b) the evaluation of risk levels; (c) implementation or planned implementation of measures aimed at controlling or mitigating those risks as low as reasonably practicable (ALARP). In addition, safety cases demonstrate that the duty holder has established response protocols that, when executed, can fulfill the legal objectives. Thus, the safety case serves as a verification tool for both the duty holder and the Health and Safety Executive (HSE). It ensures that the accepted risk-reducing measures and the safety management systems are properly established and functioning as required (HSE 2006). It should be noted that, unlike the Norwegian regulations, the safety case is primarily a tool for risk management in relation to personnel. However, this focus changed following the enactment of the Directive on Safety of Offshore Oil and Gas Operations (2013/30/EU), the scope of which has been expanded to include both the personnel and environment.

The Safety Case regulations introduce an additional level of regulatory oversight beyond other regulations, and were revised in 2015 to fully implement the EU Offshore Directive (European Commission 2013), as well as to include lessons learned from Montara and Macondo blowout accidents.

1.3.3 EU Regulations

The catastrophic Macondo blowout and explosion accident (2010) in the Gulf of Mexico raised high public awareness regarding the risks of offshore petroleum operations. In direct response to this heightened societal concern, a new directive was issued by the European Commission (2013), and requires that operators ought to:

(a) reduce the risk level of major accidents to ALARP;
(b) provide a comprehensive description of risk assessments and arrangements for major accident prevention in the report on major hazards;
(c) prepare site-specific internal emergency response plans based on the risks and hazard scenarios identified in the report on major hazards, submit these plans to the relevant authorities, and maintain the necessary resources for their prompt implementation when needed.
(d) ensure access to adequate physical, human and financial resources to prevent major accidents and mitigate their consequences.

It is evident that the EU has embraced the safety case approach as a foundation for developing its new safety directive.

1.3.4 Relevant Standards

Other relevant standards are outlined below for the readers' reference. Thus, further elaboration on these standards will not be provided.

- ISO/IEC Guide 51 (ISO 2014): Safety aspects—Guidelines for their inclusion in standards.
- ISO/IEC Guide 73 (ISO 2009): Risk management—Vocabulary.
- ISO 10418 (ISO 2019): Offshore production installations—Process safety systems.
- ISO 13702 (ISO 2015): Control and mitigation of fires and explosions on offshore production installations—Requirements and guidelines.
- ISO 15544 (ISO 2000): Offshore production installations—Requirements and guidelines for emergency response.
- ISO 17776 (ISO 2016): Offshore production installations—Major accident hazard management during the design of new installations.
- ISO 31000 (ISO 2018): Risk management—Guidelines.
- NORSOK Z-013 (Standards Norway 2023): Risk and emergency preparedness assessment.

1.3.5 Are Regulatory Approaches in Norway and the UK Suitable as Models?

A "risk-based" regulatory approach, which has long-since been implemented in the UK and Norway, was recommended for adoption into the new US regulations by the National Commission (2011) on the BP Deepwater Horizon Oil Spill and Offshore Drilling. Thus, it is meaningful to discuss the suitability of the regulatory approaches in Norway and the UK as models for other oil-producing nations.

With respect to the Norwegian regulations, there are two main principles. The first one is the unique combination of risk-based regulations, internal control and certain prescriptive regulations. Nonetheless, this combination has severe weaknesses as enterprises themselves are able to establish their own risk criteria and perform the risk assessment, without interference or approval from authorities. For instance, the regulations pay close attention to ALARP criterion, which may have been neglected by the offshore petroleum enterprises. The other main principle is the requirement of "continuous improvement" for enterprises. This indicates that the enterprises should constantly strive to minimize the risk of hazardous events. Nonetheless, this may fall short with respect to major accident prevention. While these principles have proven effective over the years, some uncertainties still exist regarding their robustness for future challenges. Certain countries in Africa and the Far East have implemented risk-based regulatory systems that bear resemblance to the Norwegian approach.

In contract, the UK regulations require the submission of a safety case, which is regarded as an improvement over the Norwegian approach by the authors, as there is a lack of a document integrating all the risk assessment arguments in the Norwegian approach. Furthermore, acceptance by the UK authorities is mandatory for all safety cases and material revisions (HSE 2006). Nonetheless, it should be noted that there are also criticisms of safety cases. Some scholars are worried that safety cases may lead to a culture of "paper safety", potentially undermining the safety in actual practice (Leveson 2011). The safety case approach has been adopted by several nations within the British Commonwealth, such as Australia, Canada, and New Zealand.

It is important to emphasize that both the Norwegian and UK approaches necessitate the presence of strong and competent regulatory authorities to conduct independent evaluations of risk studies. On balance, the UK approach is generally regarded as an easier approach for adoption by other countries, as it does not require the special combinations of diverse types of requirements in the Norwegian system.

The authors recommend that some, but not all, aspects of Norwegian and UK regulations may be used as models for other nations on the basis of their actual situations.

1.4 Uncertainty and Sensitivity

This section discusses two distinct concepts: uncertainty and sensitivity. Readers tend to confuse these two important concepts to some extent.

There are typically two categories of uncertainty: aleatory uncertainty and epistemic uncertainty. Aleatory uncertainty is irreducible uncertainty stemming from real variation of quantities. This indicates that aleatory uncertainty arises when identical experiments yield varying results under the equivalent conditions. More experiments will not reduce this variability, but provide more accurate information with respect to the probability distribution describing the variability of the results. Epistemic uncertainty is the uncertainty that may be reduced through the acquisition of knowledge. This indicates that epistemic uncertainty arises due to a lack of knowledge.

Sensitivity is defined according to the possibility of bringing forth an altered conclusion (Standards Norway 2023). The sensitivity studies could provide an effective way to evaluate the uncertainty.

1.4.1 Uncertainty: Objective or Subjective?

According to the description of risk, risk quantification necessitates the application of probability theory. There are primarily three types of interpretations of the probability, as outlined by SRA (2018).

(1) Classical probability

The classical probability framework applies exclusively to a finite sample space ($S = \{\omega_1, \omega_2, \ldots, \omega_n\}$) with equiprobable elementary outcomes. The probability measure for an event ($E \subseteq S$) is axiomatically defined as

$$P(E) = \frac{|E|}{|S|} = \frac{\text{cardinality of favorable outcomes}}{\text{total possible outcomes}} \tag{1.7}$$

(2) Frequentist probability

Within the frequentist paradigm, the probability of an event E is interpreted as "the limit of the relative frequency of the event E occurring as the number of trials approaches infinity":

$$P(E) = \lim_{n \to \infty} \frac{N_E(n)}{n} \tag{1.8}$$

Frequentist probability is an unknown quantity that necessitates estimation.

(3) Subjective probability

Subjective probability represents the analyst's personal degree of belief regarding the likelihood of an event E occurring.

Certain risk analysts claim that the classical and frequentist approaches are objective, positing that the probability is the inherent attribute of an event and exists independently of the analysts themselves. In contrast, the authors advocate for the subjective interpretation of the probability, positing that the risk value expresses uncertainties. This implies that it is crucial for the analysts to be aware of the factors affecting the extent of subjectivism in the analysis and the primary emphasis should be placed on the minimization of inherent uncertainty. Therefore, it is of importance for risk analysts to direct their attention towards the contributors to uncertainty in relation to QRA.

1.4.2 Contributors to Uncertainty

According to the QRA process, contributors to uncertainty can be systematically categorized into three primary types (Rausand and Haugen 2020).

(1) Model uncertainty

Model uncertainty consists of two dimensions: model choice and model comprehension. Model choice pertains to whether the primary characteristics of the study object are adequately mirrored or overlooked. Model comprehension correlates with the model's objectives, assumptions, limitations, input requirements, etc.

(2) Parameter uncertainty

QRA typically requires a diverse array of input data for risk modelling. Challenges arise concerning the relevance, quality, reliability, and quantity of the data. In certain instances, the available data may not be able to meet the analysis requirements.

(3) Completeness uncertainty

This type of uncertainty is primarily associated with the effectiveness of the QRA procedure and the expertise of the involved analysts. The emphasis is placed on the completeness of three core components: the background materials, the identified hazardous events, and the scenario understanding.

1.4.3 Sensitivity Studies

Typically, three types of sensitivity studies are conducted, as follows:

- Identifying the critical assumptions and parameters,
- Evaluating the effects of changes in assumptions, premises, and input data,
- Evaluating the effects of potential improvements in measures for risk reduction.

It is evident that the sensitivity studies serve as a solid foundation for evaluating the robustness of the risk model. Hence, the sensitivity studies offer an effective means to evaluate the model uncertainty and parameter uncertainty. It is worth noting that sensitivity studies have received increasing attention in recent years, particularly from regulatory authorities.

1.5 Black Swan Risk

As stated by Taleb (2007), prior to the European exploration of Australia, individuals across Eurasia held a firm belief that all swans were white, a conviction that appeared to be thoroughly substantiated by empirical evidence. In the present day, the term "Black Swan" has emerged as a compelling concept, garnering significant attention in both the scientific and industrial communities. In accordance with the original description of Taleb (2007) on black swans, three key attributes are underlined:

> First, it is an outlier, as it lies outside the realm of regular expectations, because nothing in the past can convincingly point to its possibility. Second, it carries an extreme impact. Third, in spite of its outlier status, human nature makes us concoct explanations for its occurrence after the fact, making it explainable and predictable.

Within the framework of risk management, the second attribute indicates that it is an extreme event, commonly referred to as "major accident". The term "major accident" can be interpreted as an acute accident, which immediately or subsequently causes several serious injuries, loss of human life, serious harm to the environment, and loss of substantial material assets. The third attribute is correlated with hindsight, specifically how we rationalize this extreme event after its occurrence. The first attribute, which is the most interesting part within the context of risk management, points to our inability to predict this extreme event.

Taleb's description of black swans is grounded in the scientific principle developed by Popper (1959), positing that science is incapable of proving anything, and it could merely demonstrate that a particular model or law of nature is more likely to be correct. This indicates that the accumulation of evidence, such as the observation of an increasing number of white swans, is merely able to amplify the probability that swans are exclusively white, yet it cannot conclusively prove the non-existence of black swans.

From the perspective of risk management, the concept of black swan serves as a clear warning that people ought not to hold the belief that they can predict the occurrence of everything, which can result in severe harm to personnel, environment and economic assets. Nonetheless, it should be noted that the black swan concept has been widened by some scholars (Aven 2013; Aven and Krohn 2014) to include categories of unknown unknowns, unknown knowns and known knowns, as shown in Fig. 1.4. From a practical point of view, the authors advocate for a restriction of the black swan concept to the realm of "unknown unknowns". Otherwise, the emphasis on conducting relevant and comprehensive risk assessments is missing and an excuse arises for not taking any actions. These perspectives have also been discussed by the author (Haugen and Vinnem 2015) in detail. It is crucial for us to recognize that there may be unidentified hazards and unassessed risks. Given our inability to foresee all potential occurrences, it is imperative to prepare for the unforeseen events by designing robust systems capable of handling significant deviations and effectively mitigating the impact of failures. It's not feasible to prevent the occurrence of these events because we do not know what might happen.

The Tenth Lesson will provide a case, which further verifies and illustrates how a broader interpretation of the black swan concept may result in misleading and unfortunate effects, ultimately hindering good risk management practices.

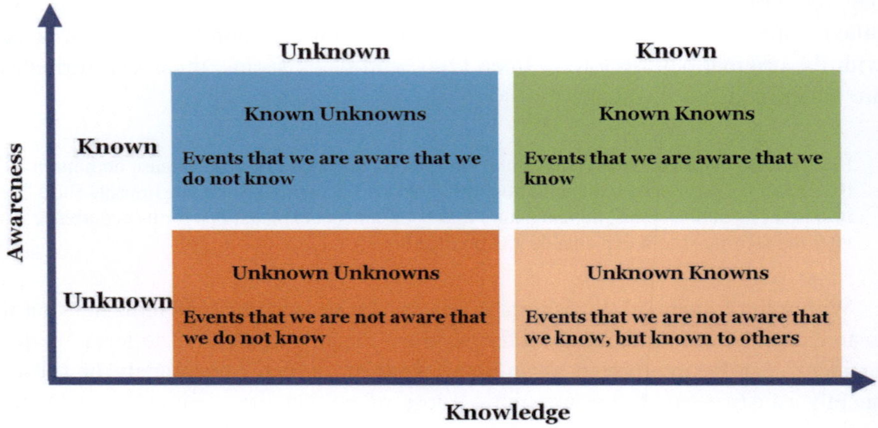

Fig. 1.4 The framework of "Known Unknowns", adapted from the statement of Rumsfeld (2002)

1.6 Influence of Human and Organizational Factors on Risk

Drawing on the lessons learned from major accidents within the offshore petroleum sector, it has been demonstrated that the development of accidents is heavily influenced by human and organizational factors (HOFs). Despite this fact, QRAs have historically concentrated on technical factors, while the concern about HOFs is missing to a large extent. Figure 1.5 presents a conceptual model for the relationship of HOFs. In response to this issue, the regulations and standards in both Norway and the UK require that QRAs address technical, organizational and human aspects.

Human factors involve individuals' perceptual, physical and mental capabilities, along with their complex interactions with workplace systems and operational environments. Additionally, human factors involve how technical system design and organizational characteristics influence human performance (Skogdalen and Vinnem 2011). Generally, there are three principal elements influencing human performance: organization, job and personal factors. Further, these elements are directly influenced by the communication system, training system, and operational procedures, all of which aim to prevent the occurrence of human errors (Stranks 2006).

Organizational factors mainly play their roles in task allocation, the design of job positions, as well as the coordination of these elements to complete activities. The organizational factors primarily encompass leadership, culture, communication, procedural quality, etc.

Over the past 15 years, significant research has focused on integrating HOFs into the QRA so as to improve risk control in operational scenarios. A systematic review has been conducted by the authors (Zhen et al. 2020) in this line of research.

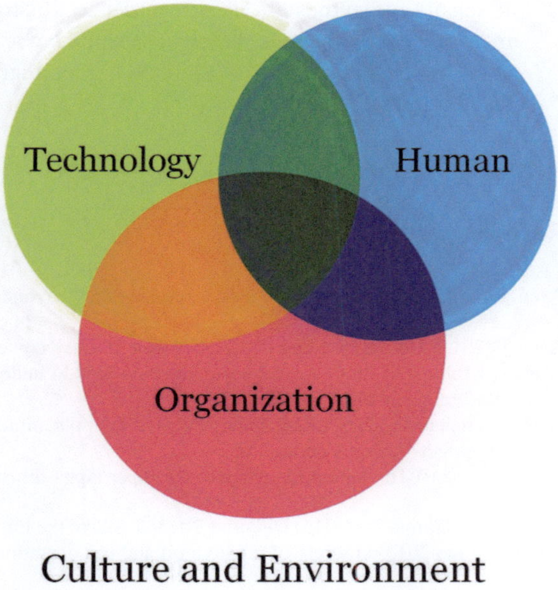

Fig. 1.5 The conceptual model for the relationship of HOFs

It can be found that though different methods have their unique characteristics, the common research strategies have been seeking answers to the following key questions:

- What HOFs are exerting influence on risk?
- In what ways do these HOFs exert influence on risk?
- To what extent do these HOFs exert influence on risk?

1.7 Questions

Q1: For the past five years, the following are the fatalities and manhours for the Norwegian offshore sector:
- Fatalities: 1.
- Manhours: 197.262 million hours.

Please calculate the average FAR value for the period using the data provided.

Q2: Based on the data in Question 1, what is the average annual PLL value for the Norwegian offshore sector?

Q3: Based on the data in Question 1, what is the average IR for the Norwegian offshore sector?

Q4: For the situation described in relation to Fig. 1.1, if we assume that you are not familiar with the local conditions around the Kjerag bolt, which uncertainty factors would dominate if you try to assess risk subjectively?

Q5: If sensitivity studies are used in order to illustrate the effect of risk-reducing measures, will the uncertainty associated with the risk results of the sensitivities be equal to, less than or higher than the risk results in the base calculation?

References

Aven T (2008) Risk analysis: assessing uncertainties beyond expected values and probabilities. Wiley

Aven T (2013) On the meaning of a black swan in a risk context. Saf Sci 57:44–51

Aven T, Krohn BS (2014) A new perspective on how to understand, assess and manage risk and the unforeseen. Reliab Eng Syst Saf 121:1–10

Aven T, Vinnem JE (2007) Risk management with application from the offshore petroleum industry. Springer-Verlag, London Ltd

Cullen L (1990) The public inquiry into the piper alpha disaster. Department of Energy. HMSO, London

European Commission (2013) Directive 2013/30/EU of the European Parliament and of the council of 12 June 2013 on safety of offshore oil and gas operations and amending directive 2004/35/EC

Haugen S, Vinnem JE (2015) Perspectives on risk and the unforeseen. Reliab Eng Syst Saf 137:1–5

Havtil (2015) Terms and expressions. Last amended in 2015

Havtil (2019) Regulations relating to health, safety and the environment in the petroleum activities and at certain onshore facilities. Last amended 26 April 2019

Havtil (2020a) Regulations relating to management and the duty to provide information in the petroleum activities and at certain onshore facilities. Last amended 16 December 2020

Havtil (2020b) Regulations relating to design and outfitting of facilities, etc. in the petroleum activities. Last amended 16 December 2020

Havtil (2020c) Regulations relating to conducting petroleum activities. Last amended 16 December 2020

HSE (2001) Reducing risks, protecting people: HSE'S decision-making process. HMSO, Norwich

HSE (2006) A guide to the offshore installations (safety case) regulations 2005. Surrey

ISO (2000) Offshore production installations—requirements and guidelines for emergency response. ISO 15544. International Standards Organization, Geneva

ISO (2009) Risk management—vocabulary. ISO/IEC guide 73. International Standards Organization, Geneva

ISO (2014) Safety aspects—guidelines for their inclusion in standards. ISO/IEC guide 51. International Standards Organization, Geneva

ISO (2015) Control and mitigation of fires and explosions on offshore production installations—requirements and guidelines. ISO 13702. International Standards Organization, Geneva

ISO (2016) Offshore production installations—major accident hazard management during the design of new installations. ISO 17776. International Standards Organization, Geneva

ISO (2018) Risk management—guidelines. ISO 31000. International Standards Organization, Geneva

ISO (2019) Offshore production installations—process safety systems. ISO 10418. International Standards Organization, Geneva

Kaplan S (1997) The words of risk analysis. Risk Anal 17(4):407–417

Kaplan S, Garrick BJ (1981) On the quantitative definition of risk. Risk Anal 1(1):11–27

Leveson N (2011) The use of safety cases in certification and regulation. ESD working paper. Massachusetts Institute of Technology

National Commission (2011) Deep water: the gulf oil disaster and the future of offshore drilling—recommendations. Report to the President. National Commission on the BP Deepwater Horizon Oil Spill and Offshore Drilling

NMA (2017) Regulations of 22 December 1993 No. 1239 on risk analyses for mobile offshore units. Last amended 19 December 2017

Popper K (1959) The logic of scientific discovery. Hutchinson & Co, London

Rausand M, Haugen S (2020) Risk assessment: theory, methods, and applications, 2nd edn. Wiley

Røyksund M, Engen OA (2020) Making sense of a new risk concept in the Norwegian petroleum regulations. Saf Sci 124:104612

Rumsfeld D (2002) Statement on Knowns and Unknowns. U.S. Department of Defense News Briefing.

Skogdalen JE, Vinnem JE (2011) Quantitative risk analysis offshore—human and organizational factors. Reliab Eng Syst Saf 96(4):468–479

SRA, 2018. Society for risk analysis glossary. Updated August 2018

Standards Norway (2023) Risk and emergency preparedness assessment. NORSOK Z-013. Rev. 4

Stranks J (2006) The manager's guide to health and safety at work. Kogan Page Ltd, London

Taleb NN (2007) The black swans—the impact of the highly improbable. Allen Lane, London

The Secretary of State (1995a) The offshore installations (prevention of fire and explosion, emergency response) regulations. Legislation.gov.uk

The Secretary of State (1995b) The offshore installations and pipeline works (management and administration) regulations. Legislation.gov.uk

The Secretary of State (1996) The offshore installations and wells (design and construction etc.) regulations. Legislation.gov.uk

The Secretary of State (2005) The offshore installations (safety case) regulations. Legislation.gov.uk

The Secretary of State (2015) The offshore installations (offshore safety directive) (safety case etc.) regulations. Legislation.gov.uk

Vinnem JE, Røed S (2020) Offshore risk assessment: principles, modelling, and applications, 4th edn. Springer-Verlag, London Ltd

Zhen XW, Vinnem JE, Yang X, Huang Y (2020) Quantitative risk modelling in the offshore petroleum industry: integration of human and organizational factors. Ships Offshore Struct 15:1–18

Second Lesson: Process and Main Elements of Risk Management

2

> "*Reasonably practicable*" *is a narrower term than* "*physically possible*" *and implies that a computation must be made by the owner in which the quantum of risk is placed on one side of scale and the sacrifice involved in the measures necessary for averting the risk* (*whether in money, time or trouble*) *is placed on the other side, and that, if it be shown that there is a gross disproportion between them—the risk being insignificant in relation to the sacrifice*—the defendants discharge the onus on them.
>
> —*Court of Appeal (1949)*

Abstract

According to the international standard ISO 31000, risk management refers to coordinated activities to direct and control an organization with regard to risk. Nowadays, there is a strong desire and enthusiasm across various industries to implement risk management in organizations. It is recognized that risk cannot be eliminated, and must be managed. This lesson unpacks the core principles of risk management, starting with the distinction between qualitative and quantitative risk assessment. Then, the lesson explores how to select risk assessment methods. The lesson further outlines the risk management process, and the main elements of risk management are introduced. Notably, the most recent NORSOK Z-013 standard is followed. A focus is placed on defining the risk tolerance criteria (RTC) and the ALARP principle, which seeks to reduce risk to a level that is as low as reasonably practicable. On this basis, the lesson examines the philosophical conflict between RTC and ALARP. Lastly, challenges in implementing quantitative risk assessment are addressed. The lesson prepares readers to grasp the fundamentals of risk management.

X. Zhen and J.-E. Vinnem, *Ten Lessons on Risks*,
https://doi.org/10.1007/978-3-031-99064-9_2

2.1 Quantitative or Qualitative Risk Assessment?

Risk assessment can be classified into two primary forms: qualitative and quantitative. Qualitative risk assessment indicates that probabilities and consequences of a risk assessment are determined by descriptive metrics, whereas quantitative risk assessment (QRA) employs numerical values to quantify the probabilities and consequences.

There has long been a debate on the choice between quantitative and qualitative risk assessments in both academia and industry. When it comes to this choice debate, the primary purpose of risk assessment should be clarified in the first place, i.e., decision-making support. Nonetheless, as has been stated in the First Lesson, this primary purpose is often forgotten.

Let's delve into this choice debate. The counter-arguments on the critics of quantitative risk assessment often argue that the quantification process is not trustworthy and the obtained numerical values are extremely uncertain. From the perspective of authors, QRA is often regarded as more precise and thus more worthy to be trusted than qualitative risk assessment. This can be attributed to the susceptibility to manipulation of the qualitative risk assessment, allowing conclusions to be shaped according to what decision-makers really want. In addition, uncertainty should not be a strong argument for the supporters of qualitative risk assessment, because both qualitative and quantitative risk assessments cannot ignore uncertainty and should evaluate it on the strength of knowledge.

In the authors' view, QRA offers significant advantages over qualitative risk assessment. For one thing, quantification improves the precision level in the support of decision-making by enabling multi-objective optimization among various factors, such as risk, cost and time. Additionally, in the consequence analyses of certain risk assessments, the accidental loads are required to be explicitly defined as design parameters that can feed into the design. Obviously, qualitative risk assessment falls short in such cases.

What is noteworthy is that the quantification or the numerical value itself is not the goal or the most important output of the risk assessment; rather, it should serve as an effective tool to support decision-making.

2.2 Selection of Risk Assessment Methods

The selection of risk assessment methods should systematically address technical constraints, organizational limitations, and operational complexities inherent to the offshore installation, while maintaining an optimal time–cost equilibrium. Following the guidance outlined in the international standard ISO 17776 (ISO 2016), the method selection should account for the following factors:

- Size and complexity of the installation,
- Identified major accident hazards,
- Potential consequence severity of the major accident,

- Level of uncertainty,
- Level of risk,
- Number of individuals exposed to risk,
- Proximity to environmentally sensitive zones.

Notably, the approaches employed in risk assessment could differ considerably based on the installation's scale and its specific phase of the project, for instance:

- For simple installations, checklists derived from previous risk assessments of similar installations and operations can be employed to provide a consistent approach for managing major accidents.
- For complex installations, a structured approach shall always be employed to ensure that no major accident hazards are overlooked. Within this structured approach, certain areas of the installation may allow for the utilization of previous risk assessment findings to minimize the necessity for new assessments.
- For new installations with a repeat of earlier design, the prior risk assessment works for the original design can be employed, subject to certain requirements.
- For installations in the early design phase, a less detailed approach can be employed.

2.3 Risk Management Process

A variety of risk management models have emerged across diverse disciplines and industries. The most prevalent model for risk management originates from the international standard ISO 31000 (ISO 2018), as illustrated in Fig. 2.1. The risk management process primarily encompasses the following steps: (a) scope, context, criteria, (b) risk assessment, (c) risk treatment, (d) communication and consultation, (e) monitoring and review and (f) recording and reporting. The risk assessment process further comprises three key sub-steps: risk identification, risk analysis, and risk evaluation. It is noteworthy that even though the whole process of risk management is often portrayed as a linear sequence of steps, it is actually iterative when implemented in practice.

The risk management framework in the ISO 31000 standard provides a generic yet systematic methodology, allowing flexible adaptation to various contexts of risk-informed decision-making within an organization, such as decisions made at strategic, operational, and project levels. The primary differences across these application areas lie in two dimensions: the defined scope and the risk analysis method employed. If the scope is defined to address risk related to a work task in the operational phase, a HAZID may be employed as the risk analysis method. However, if the scope is defined to address major accident risk, QRA is the preferred choice among risk analysis methods. This book focuses on major accident risk as the primary scope and adopts QRA as the main risk analysis method.

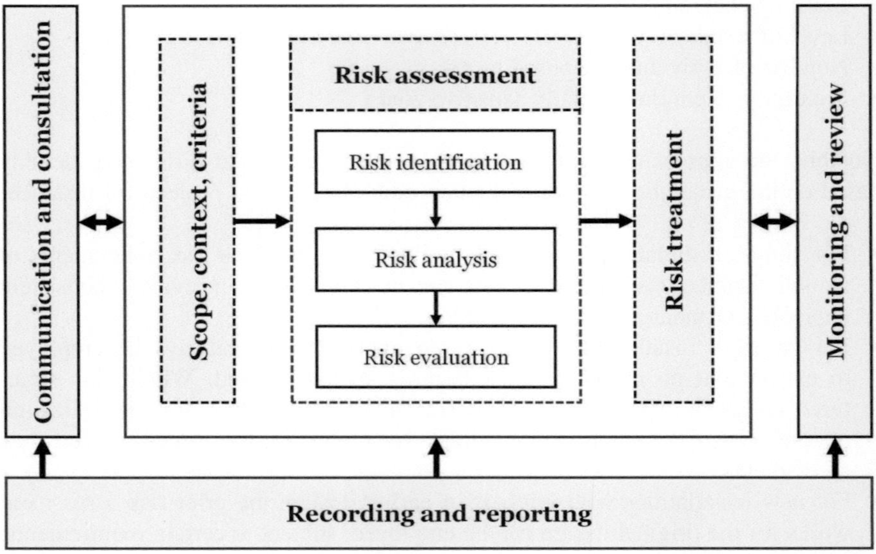

Fig. 2.1 The process for risk management. *Source* Adapted from ISO 31000 (2018)

Notably, the risk management model in the ISO 31000 standard has been incorporated into the regulatory framework and standards governing the safety of the Norwegian petroleum sector. Figures 2.2 and 2.3 illustrate the risk assessment process in the 2010 and 2023 versions of the NORSOK Z-013 standard, respectively. The relationships among risk analysis, risk evaluation and risk assessment are clearly depicted. Notably, the most recent version of the NORSOK Z-013 standard presents simplified as well as detailed methods for risk analysis. The focus of the simplified methods is on providing sufficient decision support for safe design, rather than on laying the foundation for detailed quantitative analysis of risk.

2.4 Main Elements of Risk Management

The risk management process presented in this book is established on the foundation of the ISO 31000 standard, while its risk assessment process adheres to the most recent NORSOK Z-013 standard. The risk management process can be decomposed into the following nine elements. The first six elements form the central pillars of the process, known as (a) scope, context, criteria, (b) risk assessment and (c) risk treatment, which are consistent with the established industry practice in the offshore petroleum sector. According to the NORSOK Z-013 standard (Standards Norway 2010), communication and consultation seek to assist both internal

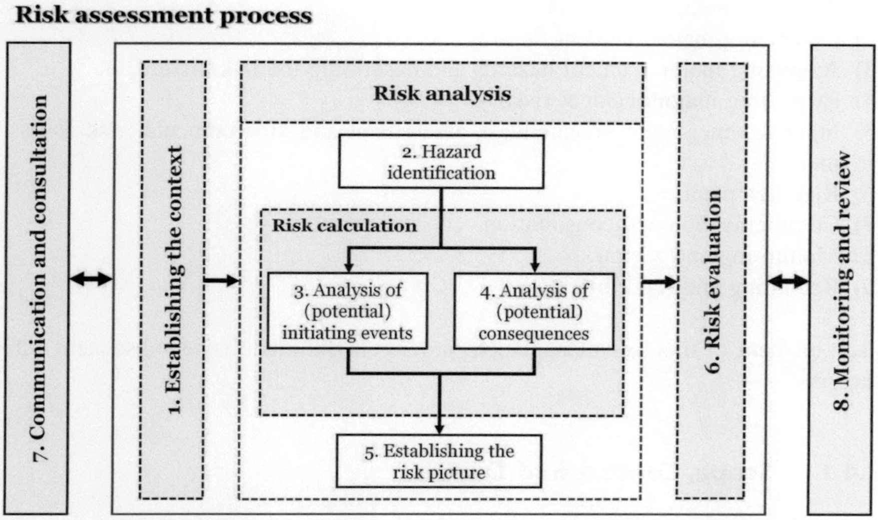

Fig. 2.2 The process for risk assessment. *Source* Adapted from NORSOK Z-013 (2010)

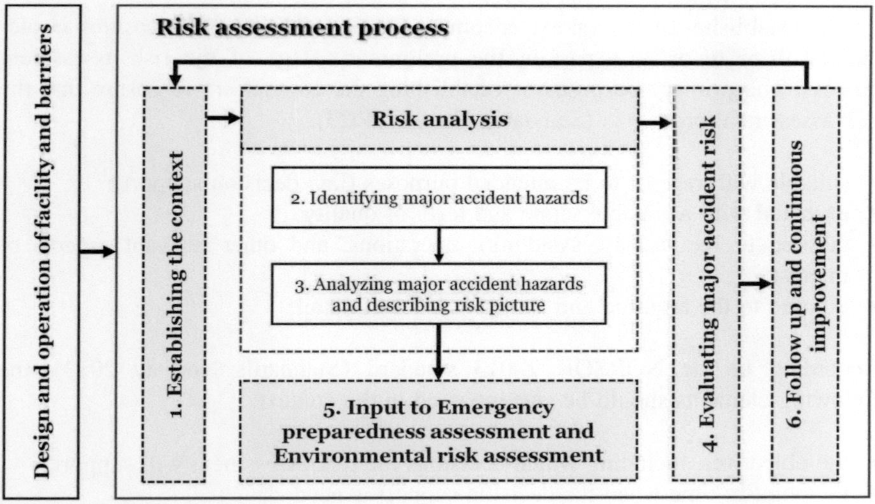

Fig. 2.3 The process for risk assessment. *Source* Adapted from NORSOK Z-013 (2023)

and external stakeholders in understanding the risk and the basis for decision-making; while monitoring and review seek to improve the future risk management process. Further, recording and reporting should be documented to provide information with respect to decision-making, communication, improvement of risk management activities, and interaction with other stakeholders.

(1) Scope, context, and criteria,
(2) Identifying major accident hazards,
(3) Analyzing major accident hazards and describing the risk picture,
(4) Evaluating major accident risk,
(5) Input to emergency preparedness assessment and environmental risk assessment,
(6) Risk treatment,
(7) Communication and consultation,
(8) Monitoring and review,
(9) Recording and reporting.

Each element of this risk management process is outlined in the subsequent subsections.

2.4.1 Scope, Context and Criteria

The risk management process is designed to be broadly applicable to any risk-based decision across different contexts. A key requirement for successful risk assessment is clearly defining its specific purpose. That is to say, it is important to know exactly which decisions the risk assessment will support.

The establishment of context encompasses all activities and measures implemented prior to or as a part of the preliminary stage of the risk assessment process. The primary purposes of establishing the context are to ensure that the risk assessment process is (Standards Norway 2023):

- suitable with respect to its intended purposes (i.e., decision support);
- executed with a suitable scope and level of quality;
- tailored to the facility, system(s), operations, and other relevant aspects of interest;
- tailored to the required and available level of detail.

According to the NORSOK Z-013 standard (Standards Norway 2023), the following elements should be encompassed in the context:

- the objectives, including which decisions the risk assessment will support,
- the point in time when the decision support is needed;
- the regulatory requirements;
- the risk criteria;
- the description of the system and the activities to be analyzed.

It should be emphasized that a deep comprehension of the system and activities to be analyzed is of importance, as this comprehension is a prerequisite for identifying and understanding the relevant hazards. It can be seen that the risk assessment process interacts with the system and activities. The description of the system and activities, along with the assumptions, forms the basis for risk assessment. Notably, the boundaries of the system (i.e., what is excluded) and any assumptions made regarding the system and activities should be clearly documented.

There is no doubt that the establishment of context is a vital step in the risk assessment process. However, there has often been a lack of adequate focus on this issue. In many instances, some authorities pointed out that the risk assessments conducted were not adequately aligned with their intended purposes. This is likely a relevant criticism of establishing the context for risk assessment.

The point is to highlight that the risk criteria ought to be established at this stage, i.e., the preliminary stage of the risk assessment process. These criteria subsequently serve as a benchmark for evaluating the risk analysis results.

2.4.2 Identifying Major Accident Hazards

It is essential to conduct a meticulous and exhaustive identification and documentation of all major accident hazards (MAHs), as any failure to identify a MAH at this stage will inevitably result in the exclusion of this hazard from further evaluations. Thus, a well-planned and comprehensive hazard identification (HAZID), which is grounded in extensive insight into design, operations and hazards, lays a solid foundation for other steps of risk assessment.

HAZID is carried out, with the aim of:

- identifying hazards related to the defined system and evaluating the sources of these hazards, as well as the events or circumstances that may give rise to their emergence and the possible consequences;
- generating a comprehensive inventory of hazards derived from the events and circumstances that could result in unwanted consequences within the scope;
- identifying possible measures for risk reduction.

In the NORSOK Z-013 standard (Standards Norway 2023), HAZAN (hazard analysis) is incorporated within the framework of HAZID. HAZAN is primarily concerned with enhancing the comprehension of MAHs and the strength of knowledge to determine if any deviations from normal occur, as well as if the simplified risk analysis method is applicable.

Here are two warnings that need to be noted. For one thing, the authorities claim that too little time and effort have often been placed on HAZID. This implies that only known hazards, i.e., known knowns and known unknowns (as indicated in Fig. 1.4), may have been identified. In this case, the HAZID fails thoroughly. Practically, identification of unknown hazards, i.e., unknown knowns and unknown unknowns (as indicated in Fig. 1.4), is much more important for preventing major accidents. For another, the inefficient use of HAZID should be minimized, as the wheel does not need to be reinvented each time. The inefficient use of HAZID manifests primarily in three distinct aspects as follows.

- A direct copy from previous HAZIDs,
- A waste of too much time on the well-known hazards,
- Lack of detailed insight into design and operation.

The authors suggest that the previously conducted HAZID works for well-known hazards can largely be reused while a tremendous amount of time and efforts should be placed on the novel or unconventional parts of the system so as to identify unknown hazards successfully.

2.4.3 Analyzing Major Accident Hazards and Describing Risk Picture

Figure 2.2 shows a classical representation of risk analysis, which includes the analysis of initiating events, analysis of consequences and establishing the risk picture. This representation aligns with the ISO 31000 standard and can be regarded as a detailed illustration focusing specifically on the risk analysis part depicted in Fig. 2.1, as well as on the analysis of major accident hazards and describing the risk picture part depicted in Fig. 2.3.

Analysis of Initiating Events

The initiating events to be analyzed are determined through the HAZID process. Analysis of initiating events serves two primary objectives:

(a) analysis and identification of potential causes of the initiating events;
(b) assessing the likelihood of the initiating events occurring.

The first objective focuses on the qualitative cause analysis, whereas the second objective centers on the quantitative analysis. Typically, the qualitative cause analysis is conducted before the quantitative analysis. However, in certain instances, the quantitative analysis may not be feasible due to a lack of available data or the absence of requirements on quantification. The importance of the qualitative cause analysis should be emphasized as it lays the foundation of accident prevention, contingent upon the ability to eliminate or control the identified causes.

When the frequency of the identified initiating events is calculated, the following requirements apply:

- Data suitability: The failure and accident data utilized must be suitable for the context of the study, aligning with the adopted methods, models, and tools.
- Cause Analysis: In the absence of representative or suitable data, an explicit analysis of potential causes for initiating events shall be conducted.
- Trends data substantiation: If trends data are employed, they should be substantiated.
- Documentation: the applied data and a discussion of their relevance should be documented.

It is noteworthy that the analysis of initiating events exhibits considerable similarities with traditional reliability analysis, which seeks to determine the probability of an item operating over a specified duration without failure. Consequently, they can employ similar techniques, for instance, the fault tree analysis, Monte Carlo simulation, Bayesian networks, among others.

Analysis of Consequences

Analysis of consequences seeks to achieve the following objectives:

(a) Assessment of the potential consequences of identified initiating events;
(b) Analysis of the possible sequences of events that could arise subsequent to the initiating event;
(c) Assessment of the impact of barrier performance, the scale of physical effects, and the degree of harm to the personnel, environment, and assets.

Among these objectives, the greatest challenge is posed by modelling the full consequences of an accident, due to the necessity of identifying a series of event nodes and accounting for the timing of each sequence.

Given the variations in the objective and scope of a risk assessment, the approaches used to analyze potential consequences can vary significantly. The approach may involve detailed modelling techniques, such as the use of extensive event trees. Alternatively, the approach can also encompass coarse assessment, which might involve extrapolation from empirical studies or existing data. Thus, consequence analysis may be qualitative or quantitative, contingent on the specific context.

Establishing the Risk Picture

Establishing the risk picture aims at formulating a clear and informative summary of the results from the risk analysis, in order to provide the necessary decision support and to facilitate evaluation against the predefined objectives and risk criteria.

Notably, the description of the risk picture is frequently confined to the parameters associated with the expression of risk criteria. This reflects a significant misconception that needs correction. As outlined in the Management Regulations of Havtil (2020), the description of the risk picture should encompass the following requirements:

- Presentation of analysis and results in a balanced and comprehensive manner, tailored to the target groups,
- Presentation of results from sensitivity analyses,
- Presentation of results from uncertainty evaluations.

The requirements outlined above are presented in a functional manner, allowing risk analysts the flexibility to interpret as needed. It is important to highlight the following points:

- Description of the risk picture should be customized to align with the specific phase of the project,
- Description of the risk picture should be adapted for various user groups,
- Description of the risk picture should provide a comprehensive input for the ALARP evaluation,
- The primary objective of establishing the risk picture is to lay the foundation for informed decision-making concerning measures for risk reduction.

Based on the authors' experience, it is advisable to incorporate the following elements so as to establish a balanced and comprehensive risk picture:

- All identified MAHs, along with their associated accident scenarios,
- The overall risk level, as well as the risk level associated with specific groups and areas within the installation,
- The contributions to risk,
- The significant improvements in the risk level,
- The sensitivity analysis pertaining to risk-reducing measures and risk model parameters,
- The uncertainty evaluation,
- An easy-to-understand format for risk description.

2.4.4 Evaluating Major Accident Risk

Risk evaluation aims at establishing a foundation for decision-making, utilizing the results of risk analysis in its context. This process primarily focuses on comparing the results of risk analysis with predetermined risk criteria in order to determine whether the risk level is acceptable, and whether additional treatment is needed.

According to the ISO 31000 standard (ISO 2018), risk evaluation can result in several decisions, including:

- Taking no additional actions,
- Considering potential risk treatment options,
- Conducting further analysis to enhance the understanding of risk,
- Maintaining the current controls,
- Reconsidering the established objectives.

What is worth being highlighted is that the decisions to be made should consider the wider context as well as the actual and perceived consequences to all stakeholders.

2.4.5 Input to Emergency Preparedness Assessment and Environmental Risk Assessment

According to the NORSOK Z-013 standard (Standards Norway 2023), the accident scenario descriptions in establishing the risk picture are input to the emergency preparedness assessment. It should be noted that not all defined situations of hazards and accidents (DSHAs) are included or analyzed in detail within a risk analysis.

In addition, the risk analysis shall provide results that serve as input to environmental risk analysis. A brief description of all identified accidental events, which could potentially result in a significant oil spills into the sea, shall be reported.

2.4.6 Risk Treatment

Risk treatment is concentrated on selecting and implementing options to manage risk. This process may be cyclical and encompasses several key steps, including:

- Preparation and implementation of risk treatment options;
- Effectiveness evaluation of that treatment options;
- Determination of the acceptability of residual risk levels;
- If residual risk levels are not acceptable, further treatment options will be taken.

The selection of the most suitable risk treatment options necessitates a careful evaluation of the benefit–cost equilibrium, required efforts, and potential disadvantages of implementation. Here are some potential risk treatment options to consider (ISO 2018):

- Eliminating the risk by opting not to initiate or persist with the risky activities;
- Taking or increasing the risk to pursue an opportunity by opting to initiate or persist with the activity;
- Removing the risk source if it is possible;
- Changing the likelihood;
- Changing the consequences;
- Sharing the risk among other parties, such as insurance, contracts;
- Retaining the risk by informed decision.

It is important to recognize that the process of risk treatment may also give rise to new risks that require management. In addition, the prioritization of risk-reducing measures was once recommended by the Norwegian Petroleum Directorate (NPD 1992) as follows.

(1) Probability reducing measures
    (a) measures reducing the probability for a hazardous situation to occur,

(b) measures reducing the probability for a hazardous situation to develop into an accidental event.

(2) Consequence reducing measures
   (a) measures relating to the design of the installation, to the load-bearing structures and passive fire protection systems,
   (b) measures relating to active safety systems, such as active fire protection systems,
   (c) measures relating to contingency equipment and contingency organization.

Although this prioritized list of risk-reducing measures is not reiterated in subsequent regulations, two important rules have been indicated. On the one hand, accident prevention is prioritized over consequence mitigation. On the other hand, the passive protection measures should supersede the active ones. The enforcement of these two critical principles during the Front-End Engineering and Design (FEED) phase of an installation has the potential to enhance its robustness in the face of major accident risk.

2.4.7 Communication and Consultation

During the risk management process, the purpose of communication and consultation is to involve pertinent internal and external stakeholders (relative to the operator) at the appropriate times and with the proper degree of participation to enhance the risk assessment's quality and ensure its adaptability and suitability to fit with intended objectives. Communication aims to foster awareness as well as comprehension of risk, while consultation focuses on collecting feedback and information to facilitate decisions. Sustaining a high level of coordination between communication and consultation is crucial.

During this process, the transfer of experience from staff possessing operational expertise is crucial for achieving a high level of safety and enhancing the predictability of risk assessment outcomes.

Clear and effective communication and consultation, both internally and externally, are essential. This ensures that all relevant stakeholders, including those potentially affected by hazards and those in charge of risk management, have a comprehensive understanding of the established context from which the risk analysis results are derived and evaluated, as well as the rationale for prioritizing specific risk treatment options.

Notably, communication and consultation often represent the fragile element during the process of risk management. This deficiency is often found in the root causes of a major accident. In particular, certain cases demonstrate that the failure of communication and consultation stems from a lack of openness and honesty, rather than the absence of action. Given this, regulations may impose specific prerequisites for the communication and consultation process.

2.4.8 Monitoring and Review

According to the NORSOK Z-013 standard (Standards Norway 2010), the process of monitoring, review, and update of the risk assessment serves several objectives:

- Monitoring the established context to ascertain its validity as a result of decisions made, emerging knowledge, or other factors that may jeopardize the validity of the context;
- Updating the context as required during the process of risk assessment;
- Ensuring that the risk assessment aligns with the updated context.

According to these objectives, monitoring and review should take place in the whole process of risk assessment, encompassing several activities:

- Analyzing and learning lessons from events, changes, and trends;
- Identifying deviations from assumptions and premises of the risk assessment;
- Identifying changes in both the external and internal context, as well as changes to the risk itself that may require revision of risk analysis and evaluation.

Notably, it is essential to update the risk analysis whenever there are any deviations from its foundational elements, including methods, models, input data, assumptions, or limitations, especially if the analysis will inform future decisions. This is because the validity of a risk analysis as a foundation for decision-making is contingent upon the verification of its foundational elements. Consequently, when updating a risk analysis, all foundational elements of risk analysis ought to be reviewed.

2.4.9 Recording and Reporting

It is essential to document and report the process for risk management along with its outcomes. The primary objectives include (ISO 2018):

- providing information for decision-making;
- communicating risk management activities and the associated outcomes across the organization;
- assisting interaction with different stakeholders;
- improving risk management activities.

In conducting a risk assessment, it is essential to meticulously document each step of the process. A well-structured report could clearly convey to stakeholders all choices made during the risk assessment process. Specific contents of a good report will encompass several key elements, such as the assumptions and scope, the decisions informed by risk analysis, the choice of risk analysis methods, the HAZID procedure, the procedure and results of risk analysis, along with the strength of

Fig. 2.4 A sample framework for organizing a risk assessment report

1. **Introduction**
   - **1.1 Objective**
   - **1.2 Scope**
   - **1.3 Limitations and Constrains**
   - **1.4 Assumptions**
   - **1.5 Abbreviations and Definitions**
2. **Description of the Study Object**
3. **Methodology**
4. **Risk Tolerance Criteria**
5. **Hazard Identification**
6. **Causal and Consequence Analysis**
7. **Risk Results and Discussions**
8. **Uncertainty Assessment**
9. **Conclusion and Risk Reducing Measures**
10. **References**

Appendix

Appendix A: HAZID Table

Appendix B: Calculations

Appendix C: Simulations

Appendix D: Data Dossier

knowledge that underpins all relevant evaluations. Crucially, the procedure and outcomes of risk management are presented in a manner that facilitates the effective comprehension of the underlying arguments for readers.

Figure 2.4 illustrates a sample framework for structuring a risk assessment report, intended to guide readers in their possible studies.

2.5 RTC—Risk Tolerance Criteria

Risk criteria are needed by the organization to achieve a balance between the goal of mitigating risk and the practical limitations. More specifically, the petroleum enterprises necessitate these criteria for the purpose of risk evaluation, which involves the evaluation of risk analysis results as well as the determination of whether additional measures for risk reduction are necessary. There are two commonly used terms pertaining to risk criteria, namely risk acceptance criteria (RAC) and risk tolerance criteria (RTC). In practice, these terms are essentially synonymous and are defined by the NORSOK standard (Standards Norway 2010) as "criteria that are used to express a risk level that is considered as the upper limit for the activity in question to be tolerable". Nonetheless, the authors prefer the term RTC over RAC in this book, as "risk tolerance" conveys the message that risk can be coped with rather than be accepted unconditionally.

As emphasized in the section on "establishing the context", RTC ought to be defined during the preliminary stage of the risk assessment process, as these criteria serve as a benchmark for evaluating the risk analysis results.

The following subsections will outline the requirements for defining the RTC and the RTC employed for personnel risk, environmental risk and economic risk.

2.5.1 Requirements in Defining the RTC

General Requirements

When defining the RTC, the following requirements should be considered:

(a) Appropriate for evaluating the relevant activities and systems;
(b) Appropriate for comparing with the results of the forthcoming risk analysis;
(c) Appropriate for supporting decision-making concerning risk-reducing measures;
(d) Appropriate for communication;
(e) Unambiguous in its definition;
(f) Not favor any specific concept or solution in the way risk is expressed.

The initial two requirements are self-evident and will not be elaborated upon. The third requirement indicates that RTC should be simple to use for evaluating the risk analysis results. Simple RTCs are typically precise and linked to specific features of the installation or activity to be analyzed, and thus suitable for evaluating the impact of risk-reducing measures. The RTC communication pertains to the interpretation and comprehension of the criteria among all relevant stakeholders, including those that are exposed to risk, the enterprise management, the regulatory authority, and the general public. Suitability for communication requires that RTC ought to be easily understandable for non-experts and enable the comparison of risks with other activities. Nonetheless, RTCs that are easily understandable may suggest an excessive simplification when addressing complex problems. RTC that enable the comparison of risks with other activities also express a societal dimension, often relating to the parameters associated with consequences, such as fatalities.

The requirement of unambiguity indicates that RTC ought to be defined in a way that prevents the emergence of unreasonable or unintended effects when risk analysis results are expressed or evaluated. The ambiguity problems may stem from issues including the imprecise formulation of RTC and the lack of clarity in defining the system boundaries. The last requirement reminds that the utilization of RTC in risk evaluation typically necessitates the prioritization of one concept over others as a result of the minimal risk. Thus, when different RCTs are defined for the same type of risk, such as personnel risk as defined by FAR compared to impairment risk for main safety functions, it can lead to discrepancies in risk prioritization.

Norwegian Regulatory Requirements

Petroleum enterprises are required to establish their own RTC in adherence to Norwegian regulations. These requirements with respect to the RTC are presented in the management regulations of Havtil and the NORSOK Z-013 standard. The following requirements need to be considered:

- The RTC should address all stakeholders who are subject to the specific risk exposure;
- The RTC should be established for both the personnel and the environment;
- Further measures for risk reduction should always be taken into account, even when the risk level falls within the RTC.

Currently, Havtil has not released any specifications or guidelines on how to establish the RTC. However, Havtil plans to issue a memo in 2026 that will outline the guidelines on the establishment and use of RTC in response to challenges encountered by petroleum enterprises in establishing their own RTC.

UK Regulatory Requirements
In the UK, the regulatory requirements pertaining to the RTC are outlined in the Safety Case Regulations (SCR), and can be summarized as follows.

- Emphasis is placed on the ALARP principle, in relation to the personnel safety and TR integrity.
- The impairment frequency of TR integrity is less than 10^{-3} per year.

It is important to note that the maximum allowable frequency for the loss of TR integrity serves as the sole quantitative criterion in the SCR. Due to this reason, the preliminary work of the Safety Case prioritizes adherence to this frequency threshold. After that, operators will focus on the personnel risk more directly.

2.5.2 RTC for Personnel Risk

Based on current industry practices, the following three types of RTC are most commonly applied for personnel risk in the petroleum sector.

(a) FAR,
(b) AIR,
(c) F-N curve.

In Norway, it has been noted that all these three approaches are employed within the petroleum sector. Many petroleum enterprises have defined a FAR of 10 as a risk tolerance limit for the average exposed individuals, while a FAR of 20 is assigned to the individuals exposed to the highest level of risk. Additionally, a significant risk tolerance limit is represented by an AIR of 10^{-3} per year, which is adopted by some petroleum enterprises for the highest risk level. Furthermore, one particular petroleum enterprise utilizes a F-N curve to establish its risk tolerance limit. The annual frequency of N fatalities for an offshore installation can be expressed by Eq. (2.1). It can be seen that the curve is adjusted in accordance with the maximum annual number of personnel on board. Note that a relatively high aversion factor can also be used in a F-N curve for defining the risk tolerance limit. Equation (2.2) presents an example, where the factor b is the aversion factor.

$$f(N) = \frac{0.01}{N \times \sqrt{\frac{\mathrm{POB}_{\mathrm{max}}}{100}}}, \quad N = \mathrm{POB} \tag{2.1}$$

$$f(N) = \frac{0.01}{N^b}, \quad N = \mathrm{POB} \tag{2.2}$$

In the UK, the risk tolerance limit set by the regulatory authority is defined by an AIR of 10^{-3} per year for existing installations, while for new installations, the limit is reduced by a factor of 10.

For the purpose of comparison, FAR=10 is taken as a benchmark. In contrast, Equation (2.2) (where $b = 1$) corresponds to FAR $= 5.9$ while AIR $= 10^{-3}$ per year corresponds to FAR $= 34$. Note that FAR $= 34$ reflects 3 persons per position, i.e., the Norwegian working schedule, while most of other countries have 2 persons per position. Thus, this suggests that the risk tolerance limit with respect to the F–N curve is the most challenging. Further, the risk distribution (instead of a single value) as the risk tolerance limit is more challenging to meet.

What needs to be emphasized is that FAR $= 10$ is a very relaxed risk tolerance limit now, as the current statistical values are well below it for worldwide operations. This relaxation reduces the practical effectiveness of the RTC in improving risk levels. The authors suggest that the regulatory authorities, rather than the industry, should bear the responsibility for establishing the stricter tolerance limits of FAR.

2.5.3 RTC for Environmental Risk

The RTC for environmental risk can be categorized into three distinct tiers:

(a) field-specific criteria,
(b) installation-specific criteria,
(c) operation-specific criteria.

It can be noted that the risk tolerance limits are quantitatively different among these three levels. When it is not obvious on which level the risk tolerance limit should be applied, it is advisable to utilize the highest level. In other words, the field-specific criteria should be prioritized, followed by the installation-specific criteria, and finally the operation-specific criteria.

The environmental RTC should encompass the frequency of pollutant discharges into the environment, which leads to specific consequences. To simplify

Table 2.1 Example of the quantitative environmental RTC (DNV GL 2018)

Damage	Recovery time	Risk tolerance limit (annual frequencies)
Minor	1 month–1 year	$<1.0 \times 10^{-3}$
Moderate	1 year–3 years	$<2.5 \times 10^{-4}$
Significant	3 years–10 years	$<1.0 \times 10^{-4}$
Serious	More than 10 years	$<2.5 \times 10^{-5}$

this process, the frequency and volume of pollutant discharges into the environment, along with their potential consequences, may be applied. Environmental consequences could refer to the recovery time of the vulnerable habitats or populations, as well as the exposure of specific areas or volumes with a certain sensitivity, such as the length of the contaminated shoreline. The environmental RTC commonly employed for offshore activities on the Norwegian Continental Shelf (NCS) is predicated on the recovery time for sensitive environmental resources.

Table 2.1 presents a typical example with respect to the quantitative environmental RTC.

2.5.4 RTC for Economic Risk

The duration of production delays is often used to express the risk to economic assets. The frequencies of events that cause delays are typically determined for the following delay categories.

- 1–7 days,
- 1 week–3 months,
- 3 months–1 year,
- 1 year–2 years,
- More than 2 years.

Notably, it is not common practice to define an explicit RTC for economic risk. Instead, the economic risk analysis is conducted first, followed by the implementation of measures to reduce risk based on the cost–benefit analysis.

2.6 ALARP—As Low as Reasonably Practicable

The ALARP principle is derived from the UK legislation enacted by the UK Parliament in 1974 (UK Parliament 1974), which mandates that risks to employees and other individuals must be reduced "so far as is reasonably practicable (SFAIRP)". The concepts pertaining to ALARP and SFAIRP are synonymous,

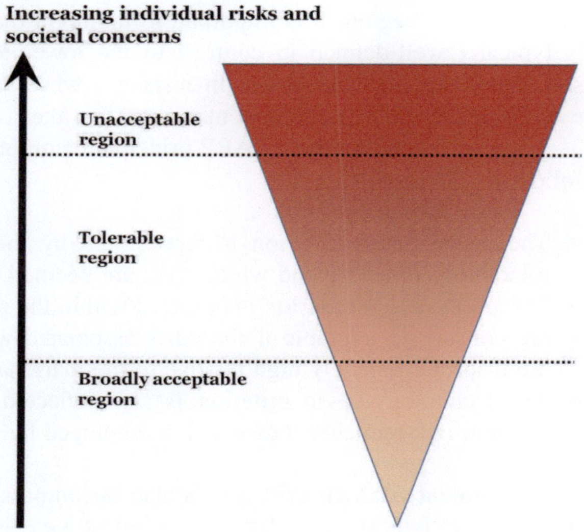

Fig. 2.5 The ALARP principle. *Source* Adapted from UK HSE (HSE 2001)

both emphasizing the principle of "reasonably practicable", which involves weighing the risks against the costs required to mitigate those risks further, alongside making decisions by good practice.

Figure 2.5 illustrates the ALARP principle as outlined by the UK Health and Safety Executive (HSE 2001). The risk level can be classified into three distinct regions:

(a) An unacceptable region: risks categorized within this region are deemed unacceptable except under extraordinary circumstances. Activities or practices that result in risks within this region are prohibited unless corresponding measures for risk reduction enable the reduced risk level falling within the following regions, or there are compelling justifications for the continuation of the activity or practice.
(b) A tolerable region, also referred to as the ALARP region: risks categorized within this region are deemed tolerable only when the costs related to risk reduction are grossly disproportionate to the benefits derived from reducing those risks.
(c) A broadly acceptable region: risks categorized within this region are typically viewed as insignificant and sufficiently controlled. Further measures for risk reduction are usually not required unless such measures are reasonably practicable.

To distinguish between the unacceptable, tolerable, and broadly acceptable risk regions, it is essential to establish two risk limits: the upper tolerability limit, which separates the "tolerable" region from the "unacceptable" region, and the lower tolerability limit, which delineates the "broadly acceptable" region from the "tolerable" region. It is important to highlight that the upper tolerability limit is typically well-defined in contrast to the lower tolerability limit, which may sometimes remain unspecified. In instances where no lower tolerability limit is established, it is necessary to demonstrate that the risk level shall be ALARP.

It can be noted that the ALARP principle incorporates three distinct criteria, as follows.

- The equity-based criterion is represented by the establishment of an upper tolerability limit, beyond which risks are deemed unacceptable.
- The utility-based criterion is reflected within the ALARP region, wherein risks are considered tolerable if the costs associated with further control measures are disproportionately high relative to the marginal reduction in risk.
- The technology-based criterion is also reflected within the ALARP region, wherein risk-reducing measures are employed by the use of good practice.

The equivalent ALARP principle is also encompassed in the Norwegian Framework Regulations (Havtil 2019), as demonstrated in Sect. 1.3.1. A visual representation of the ALARP principle as outlined in the NORSOK Z-013 standard is shown in Fig. 2.6.

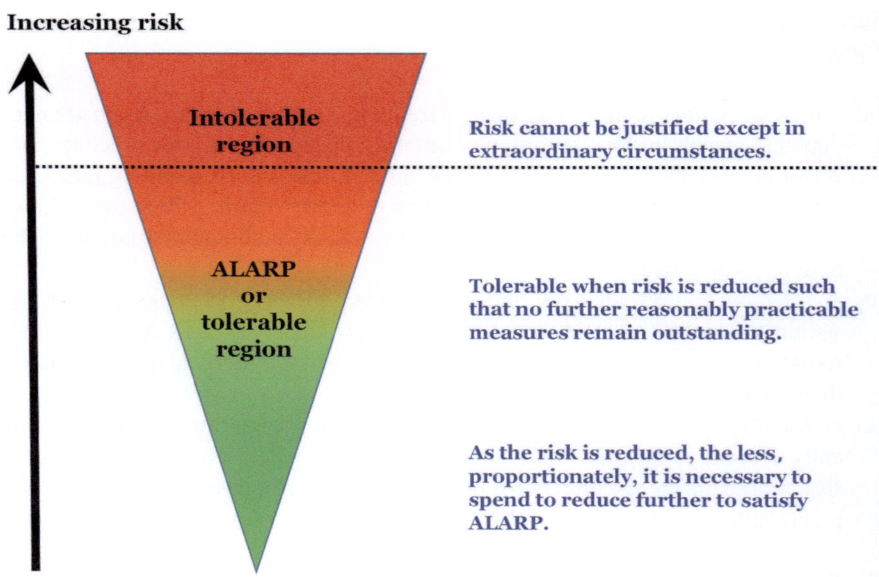

Fig. 2.6 The ALARP principle. *Source* Adapted from NORSOK Z-013 (2010)

2.7 Philosophical Dilemma Between Risk Tolerance Criteria and ALARP

The first introduction of RTC into the Norwegian offshore legislation occurred implicitly around 1980, with extensive application commencing approximately in 1990. Since that time, RTC has served as the primary tool for risk evaluation. Nonetheless, since 2004, a notable shift has taken place towards emphasizing the ALARP principle as a complementary approach to RTC.

The authors argue that a purely mechanical approach to risk evaluation, which relies solely on RTC, ought to be expanded by an alternative evaluation approach similar to ALARP. The authors posit that, within the highly mature and safety-conscious offshore petroleum industry, RTC could potentially be removed as a regulatory requirement. Notably, the regulatory framework conditions are changing in some countries, where climate issues are increasingly taking precedence over health, environment and safety issues. As a result, safety consciousness may suffer. In this situation, the authors recommend that regulations should continue to favor the inclusion of requirements for the formulation and application of RTC.

2.8 QRA Challenges

In the context of practical QRA studies, various challenges have been identified and will be succinctly addressed in the following. Note that most of these challenges are disclosed across nearly all QRA studies. However, the ethical challenges, which particularly arise from the most deficient QRA studies, will be examined in greater detail in the Tenth Lesson of this book.

(a) HAZID challenges

Firstly, the completeness of HAZID has been suspicious, as it is frequently conducted in an unsystematic manner. In particular, operational and organizational failures are often overlooked during the HAZID process. Secondly, the criteria employed to exclude non-critical hazards and hazardous events are frequently not well documented. Consequently, these non-critical hazards and hazardous events, which will not be further analyzed, lack proper documentation.

(b) Data challenges

Data serve as the core input for QRA studies, and their availability, quality, and completeness are crucial to the QRA validity. Historical accident and incident data, equipment reliability data, hazard data, human error data, and consequence data all form the essential foundation for quantitative risk modelling. However, the statistical basis of risk analysis data remains weak globally, with many datasets being incomplete or missing, a problem particularly pronounced in early-stage oil and gas facilities. Furthermore, due to the sensitive and specialized characteristics of risk analysis data, the development of effective cross-organizational data-sharing mechanisms remains a major challenge.

(c) Challenges in risk analysis

Firstly, the documentation of assumptions and premises is generally insufficient, leading to a lack of transparency during the risk analysis process. It is worth noting that unnecessary assumptions may arise when the study team lacks familiarity with the study object or fails to clarify the uncertainties. Secondly, the criteria for selecting a representative set of accident scenarios for detailed analysis are not clearly defined, which impacts the representativeness of the risk picture. Thirdly, it is found that the studies sometimes have unclear definitions of essential parameters, for instance the definition of "Group Individual Risk". It may be unclear for a staff who belongs to different groups. Sometimes this leads to errors in the calculation of risk contributions. Fourthly, it is challenging for current risk models to reflect HOFs, the importance of which has been confirmed by numerous major accidents. Fifthly, the modelling of safety barrier performance is insufficient, and the analysis of safety barriers lacks consideration of failure mechanisms, thereby neglecting dependencies and common cause failures. Lastly, comprehensive quantitative risk assessments are typically resource-intensive and time-consuming, often leading to conflicts with constraints on manpower, financial resources, and time. This challenge is particularly pronounced for small- and medium-sized enterprises, for which comprehensive quantitative risk assessments are often difficult to implement.

(d) Challenges in describing the risk picture

The results presentation of risk analysis is often unbalanced as well as lacks comprehensiveness. Particularly, the results presentation should not be restricted to RTC requirements. Additionally, the results presentation in relation to uncertainty within the petroleum sector is insufficient.

(e) Challenges in identifying measures for risk reduction

The primary challenges encountered include incomplete HAZID, incomplete risk modelling concerning HOFs, and an excessive concern with adherence to RTC.

(f) Challenges in using the risk picture for decision-making

The risk assessment does not provide adequate decision support in certain situations, where the risk assessment is conducted merely to meet the regulatory requirements. This challenge is illustrated in Fig. 2.7.

(g) Ethical challenges

It can be anticipated that ethical failures in some of the worst QRA studies are able to severely destroy the major process of QRA. Consequently, the practical value and significance of QRA results in informing decision-making will be lost.

Fig. 2.7 Challenge on using risk assessment to support decision-making

2.9 Questions

Q1: What are the potential drawbacks of quantitative risk assessment?

Q2: Could there be situations where you would want to carry out quantitative risk assessments also for a simple offshore installation?

Q3: What could be the main problem if you fail to establish the context?

Q4: If you carry out Hazard Identification without experienced personnel, what could be the critical problems?

Q5: Frequency of initiating events may be established from incident statistics or comprehensive quantitative reliability analysis, the latter is time consuming, the former is considerably less time consuming. What is the main disadvantage of using incident statistics?

Q6: The most comprehensive consequence analysis methods involve computational fluid dynamics (CFD), which is a powerful, yet very time-consuming tool. If you are using the risk assessment to establish design loads, would you prefer a coarse analysis or CFD?

Q7: What is the role of ALARP in risk evaluation?

Q8: Is seeking approval from the authorities a justified risk treatment approach?

Q9: What would you expect to provide the most risk reduction, RTC or ALARP, from a theoretical as well as a practical point of view?

Q10: How can you avoid the QRA challenges listed in Sect. 2.8?

References

DNV GL (2018) Environmental risk and emergency preparedness analysis (MRABA) for exploration wells 16/1-31 S & A in PL 338 and 16/5-8 A in PL 815 in the North Sea for Lundin Norway AS, report 2018-1122

Havtil (2019) Regulations relating to health, safety and the environment in the petroleum activities and at certain onshore facilities. Last amended 26 Apr 2019

Havtil (2020) Regulations relating to management and the duty to provide information in the petroleum activities and at certain onshore facilities. Last amended 16 Dec 2020

HSE (2001) Reducing risks, protecting people: HSE's decision-making process. HMSO, Norwich

ISO (2016) Offshore production installations—major accident hazard management during the design of new installations. ISO 17776. International Standards Organization, Geneva

ISO (2018) Risk management—guidelines. ISO 31000. International Standards Organization, Geneva

NPD (1992) Regulations relating to explosion and fire protection of installations in the petroleum activities. Stavanger, Norway

Standards Norway (2010) Risk and emergency preparedness assessment. NORSOK Z-013. Rev.3

Standards Norway (2023) Risk and emergency preparedness assessment. NORSOK Z-013. Rev.4

UK Parliament (1974) Health and safety at work etc. Act 1974

Third Lesson: Accident Lessons and Investigations

3

......, the later generations lament Qin dynasty, but do not learn lessons. Thus, later generations lament the later generations again and again.

—Mu Du (825)

Abstract

There has been a saying that "If you think accident prevention is expensive, try an accident". Some enterprises have learned this lesson the hard way. Undoubtedly, investigating accidents and learning their lessons are crucial for preventing future analogous accidents. This lesson explores the importance of learning lessons from past accidents. Then, the lesson presents and discusses the critical implications of four notable accidents across different countries. A core focus is placed on the importance of root causes, which are the most essential outcomes derived from accident investigation. The lesson further outlines the accident investigation process and compares two types of investigation, i.e., enterprise internal investigation and external investigation. After that, two commonly used investigation modelling techniques based on sequential models are introduced. Further, the lesson presents the integration principle between accident investigation and QRA so as to enhance the understanding of the accident scenario and establish a comprehensive risk picture. Lastly, the challenges of lessons learned from accidents are addressed. Some reputable accident investigation organizations are also introduced for reference. This lesson prepares readers to realize that accident lessons are not about blame but learning, ensuring errors of the past pave the way for a safer future.

X. Zhen and J.-E. Vinnem, *Ten Lessons on Risks*,
https://doi.org/10.1007/978-3-031-99064-9_3

3.1 Importance of Learning from Accidents

Major accidents have occurred uninterruptedly within the offshore petroleum sector, including the most noticeable capsizing of the Semi-submersible rig Alexander L. Kielland in Norway in 1980 (NOU 1981), the explosion and fire at the steel jacket platform Piper Alpha in the UK in 1988 (Cullen 1990), the Macondo well blowout in the United States (US) in 2010 (Graham et al. 2011), the Penglai 19-3 oil spills in China in 2011 (SOA 2012), and the gas explosion of the Floating Production, Storage and Offloading (FPSO) unit Cidade de São Mateus (CDSM) in Brazil in 2015 (ANP 2016). Table 3.1 provides a summary of major offshore accidents, serving as a reminder that "*Do not think because an accident hasn't happened to you that it can't happen*". It is worth noting that certain positive changes or improvements have arisen following some of these major accidents, such as the Alexander L. Kielland capsizing and the Piper Alpha explosion and fire. Empirical evidence confirms that the recurrence of certain analogous accidents can largely be attributed to the failure of learning lessons from prior accidents. This failure hinders the comprehension of the accident mechanism as well as the formulation of effective accident prevention and control strategies (Khan and Abbasi 1999).

Systematic learning activities from accidents encompass the following purposes:

- Collecting pertinent information concerning the accidents (*what, where, when*);
- Analyzing the causes and influencing factors associated with the accidents, along with the sequences involved (*why, how*);
- Extracting, analyzing, and communicating knowledge from the accidents (*learning*);
- Recommending and implementing preventive and protective strategies (*learning*);
- Recording and communicating the acquired knowledge to reduce the probability of analogous accidents occurring in the future (*learning*).

Regrettably, the global offshore petroleum industry exhibits a significant reluctance to disclose detailed information regarding the accidents and tends to underplay or even conceal its mistakes. This assertion is largely substantiated by the fact that the accident investigation reports are rarely made accessible to the public. In this situation, offshore petroleum enterprises should be aware that learning lessons from accidents based on the available investigation reports is particularly precious and crucial for the enterprises' safety management.

Table 3.1 Typical major accidents in the offshore petroleum sector (1975–2024)

Area	Time	Consequences
Norway	01 November 1975	A hydrocarbon leak from a ruptured test riser occurred on the steel jacket wellhead platform Ekofisk alpha, resulting in an explosion and fire associated with 3 fatalities
Norway	01 March 1976	The Deep Sea driller capsized, resulting in 6 fatalities
Norway	23 April 1977	A blowout occurred on the steel jacket wellhead platform Ekofisk bravo, resulting in a major oil spill of 20,000 m^3
US	03 June 1979	The IXTOC I exploration well experienced a blowout, resulting in a major oil spill of approximately 3.5224 million barrels
China	25 November 1979	The Bohai 2 jack-up rig capsized, resulting in 72 fatalities
Norway	27 March 1980	The semi-submersible flotel Alexander L. Kielland capsized, resulting in 123 fatalities
Canada	15 February 1982	The semi-submersible drilling unit ocean ranger capsized, resulting in 84 fatalities
China	25 October 1983	The drill ship Glomar Java Sea capsized, resulting in 81 fatalities
Brazil	16 August 1984	The fixed jacket production platform Enchova 1 suffered a blowout and fire, resulting in 42 fatalities
Norway	06 November 1985	The semi-submersible drilling unit west vanguard suffered a shallow gas blowout, resulting in an explosion and fire associated with 1 fatality
UK	06 July 1988	A gas leak from a blind flange caused an explosion and fire on the steel jacket platform piper alpha, resulting in 166 fatalities
UK	22 September 1988	A subsea blowout caused an explosion and fire on the semi-submersible drilling rig ocean odyssey, resulting in 1 fatality
US	23 March 1989	The tanker Exxon Valdez ran aground, resulting in a major oil spill of 37,000 tons
Thailand	03 November 1989	The drill ship Seacrest capsized, resulting in 91 fatalities
Brazil	15 March 2001	The semi-submersible platform Petrobras P-36 capsized, resulting in 11 fatalities
Norway	20 August 2004	The gas export pipeline from the FPSO Jotun A ruptured, resulting in a major gas release of approximately 1.3 million standard cubic meters
India	27 July 2005	A vessel collided with the offshore Mumbai high north platform, resulting in a gas leak and fire associated with 22 fatalities
Australia	21 August 2009	A blowout occurred on the Montara wellhead platform, leading to a major oil spill estimated between 4000 and 30,000 tons

(continued)

Table 3.1 (continued)

Area	Time	Consequences
US	20 April 2010	A blowout occurred on the semi-submersible rig Deepwater Horizon (DWH) in the Gulf of Mexico (GoM), resulting in an explosion and fire associated with 11 fatalities, along with a major oil spill of approximately five million barrels
China	04 June 2011 12 June 2011	Leaks took place at two wellhead platforms respectively in the Penglai (PL) 19-3 oilfield, causing severe environmental damage over an area of approximately 6200 km^2
Brazil	11 February 2015	An explosion took place on the FPSO unit CDSM, resulting in 9 fatalities
Mexico	01 April 2015	Explosion and fire took place on the Abkatun A platform, resulting in 7 fatalities
Azerbaijan	04 December 2015	A fire broke out on the Gunashli 10 oil platform, resulting in 33 fatalities
Mexico	06 February 2016	A fire broke out on the Abkatun A platform, resulting in 3 fatalities
Malaysia	27 October 2020	The Dayang topaz vessel struck the Baram B oil platform, resulting in 2 fatalities
China	05 April 2021	A fire broke out on the V platform in the PL 19-3 field, resulting in 3 fatalities
Mexico	02 July 2021	A significant blaze resembling a large "eye of fire" occurred near an oil platform due to a gas leak from an underwater pipeline
Mexico	22 August 2021	A fire broke out on the oil platform E-KU-A2, resulting in 7 fatalities
Nigeria	02 February 2022	Fire and explosion occurred on the FPSO trinity Spirit, resulting in 10 fatalities
Baltic Sea	26 September 2022	Explosions and leaks were reported in the underwater Nord stream gas pipelines, with a worst-case scenario estimating the potential release of approximately 778 million standard cubic meters of methane gas
Malaysia	01 May 2023	An explosion occurred on the oil tanker Pablo, resulting in 3 fatalities
Mexico	07 July 2023	An explosion and fire took place on the Pemex natural gas platform, resulting in 3 fatalities
Gabon	20 March 2024	A fire took place at the Perenco oil platform, resulting in 5 fatalities and 2 injuries
Mexico	06 April 2024	A fire occurred on the Pemex oil platform, resulting in 1 fatality and 13 injuries

3.2 Lessons from Oil Spills in the US GoM and Chinese Bohai Bay

This section seeks to analyze and elucidate the critical implications related to the prevention and mitigation of offshore oil spills, drawing upon two notable accidents: the Chinese PL 19-3 oil spills and the US DWH oil spill. It is important to recognize that these two major accidents share a lot of root causes in relation to organizational factors. This fact emphasizes the need and importance to learn from previous accidents once more.

3.2.1 Chinese Penglai 19-3 Oil Spills

Background

The PL 19-3 oil field, the proven petroleum reserves of which exceed 1 billion tons, represents the largest offshore petroleum field ever discovered in China. The field is located in Block 11/05 in the central south of the Bohai Sea, characterized by an average water depth of approximately 30 m. There are a total of seven production platforms (A/B/C/D/E/F/M), associated with 193 production wells, 53 water injection wells, and 6 debris re-injection wells. In 2010, the total output of petroleum reached 7.78 million tons.

Pursuant to the Production Share Contract (PSC) executed between China National Offshore Oil Corporation (CNOOC) and ConocoPhillips China (COPC) for the Block 11/05, CNOOC holds a 51% interest while COPC retains the remaining 49%. A Joint Management Committee (JMC) was established to review and approve the important issues pertaining to the field development. Furthermore, the operatorship was officially transferred from COPC to CNOOC on 1st July 2014.

In June 2011, two significant oil spill accidents took place one after the other at platforms B and C, which were put into production in May 2008 and July 2007, respectively. In accordance with the official investigation report published by the State Oceanic Administration (SOA 2012), the oil spills resulted in the contamination of an area approximately measuring 6200 km^2, as illustrated in Fig. 3.1. Notably, the Bohai Sea, located in China, is classified as a semi-closed inland sea, characterized by significantly limited self-purification capabilities. Therefore, its fragile ecosystem, along with its vital economic attributes, renders it particularly vulnerable to the adverse effects of oil spills. Ultimately, the financial compensation required for the damage inflicted on marine ecology and fishery resources amounted to 1.683 billion RMB and 1.35 billion RMB, respectively. The PL 19-3 oil spills are recognized as the most severe offshore oil spill accidents documented in the history of China.

Notably, following a period of approximately 10 years, the PL 19-3 oil field encountered a significant blowout and fire accident on April 5, 2021, resulting in three crew members being reported missing. As of now, the investigation report pertaining to this major accident is still unavailable to the public. Consequently, it is imperative to look back at the PL 19-3 oil spill accidents in order to identify potential

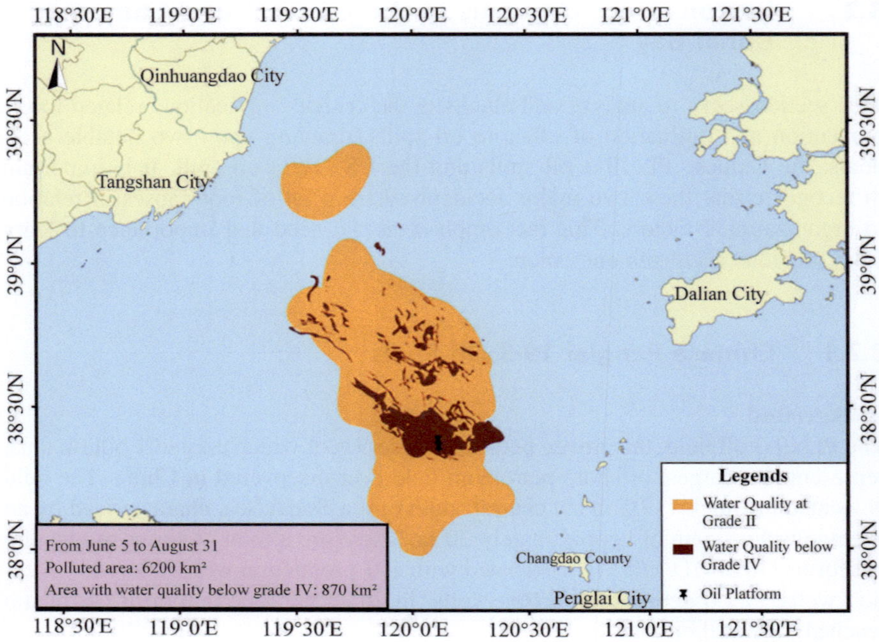

Fig. 3.1 The ranges of polluted marine areas resulting from the PL 19-3 oil spills

common root causes, thereby facilitating the prevention of future accidents in this area.

Sequence of Events

The timeline of the PL 19-3 oil spill accidents is illustrated in Fig. 3.2, encompassing the duration from the first observation of an oil spill to the complete cessation of operations in the PL 19-3 field, a period extending over three months.

On June 4, 2011, at approximately 7:00 p.m., the COPC notified the SOA regarding the observation of a minor oil presence on the sea surface in the northeastern vicinity of the Wellhead Platform B (referred to as B-Accident). In response, the SOA promptly instructed COPC to control the oil spill and undertake the self-inspection procedure. Concurrently, the SOA commenced emergency monitoring activities for the offshore oil spill. On 8 June 2011, the COPC contacted the SOA once more to report the discovery of a subsea oil spill in the northeastern area of the Wellhead Platform B. By 12 June 2011, it was confirmed through oil fingerprint identification techniques that the source of the spilled oil was the PL 19-3 oil field. Further investigation revealed that the leaks originated from a pre-existing geological fault, which had been compromised as a result of excessive pressure in relation to water injection activities. Consequently, the associated injection well B23 was closed and the pressure was relieved through backflow operations at the Wellhead Platform B.

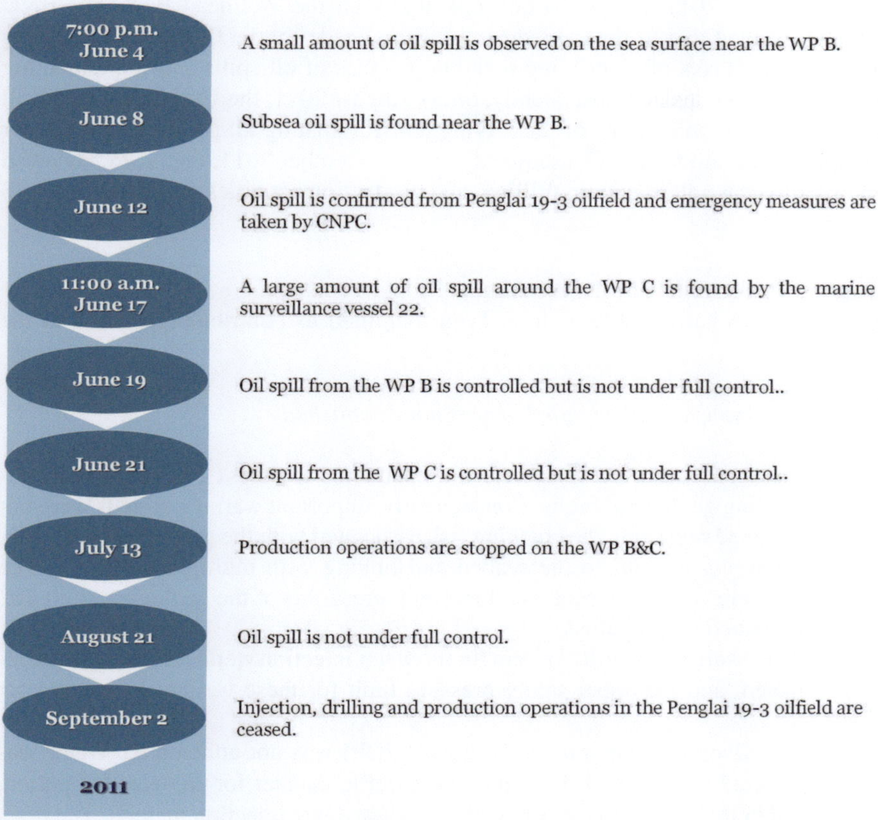

Fig. 3.2 Timeline for the PL 19-3 oil spill accidents

By 19 June 2011, the oil spill from the Wellhead Platform B had been controlled to a large extent but not under full control.

On 17 June 2011, at 11:00 a.m., a significant oil spill was observed in the vicinity of the Wellhead Platform C by the marine surveillance vessel 22 (referred to as C-Accident). Subsequent investigations revealed that the well C20 at the Wellhead Platform C had encountered a kick due to drilling into an unanticipated over-pressure layer. Although the well C20 was promptly shut in, the resulting pressure led to the leakage of crude oil and oil-based mud into the marine environment. A cementing operation was successfully executed within 48 h to halt the release, and the well was subsequently stabilized, plugged, and abandoned. By 21 June 2011, the oil spill originating from the Wellhead Platform C had been mostly contained; however, it had not been fully brought under control. It is important to note that though two consecutive subsea oil spill accidents occurred within a two-week period in the same oilfield, no direct correlation was found between the B-Accident and the C-Accident.

On 13 July 2011, oil production operations on the Wellhead Platforms B&C were suspended due to the inadequate progress made by the COPC in addressing the necessary tasks of identifying potential sources of oil spills and implementing measures to seal leaks. Subsequently, on 31 August 2011, the COPC did not fulfill the supervisory obligations of identifying and eliminating all potential sources for oil spills once and for all. Consequently, on 2 September 2011, the COPC received an order to cease all injection, drilling, and production activities within the PL 19-3 oil field.

Barrier Performance in the Accidents

The following barrier failures have been identified as contributing factors to the B-Accident:

- *The well integrity either failed or was not established.*

  (a) Pressure-measuring tests were not conducted to assess the stability of pre-existing geological faults. Consequently, important warning signals were not obtained regarding the cracking risk associated with the geological fault 502, extending upward to the seabed and linking with multiple oil layers. The pressure-bearing capacity and fracture thresholds of the geological fault 502 remained unquantified.

  (b) A pressure monitoring system for the water injection wellheads was not established, and the upper safety pressure limit for these wellheads was not set up.

  (c) The Overall Development Program (ODP) was not adhered to. Water was injected into the well B23 in a non-specific manner for an extended period, rather than utilizing the designated separate layer injection method. The pressure variations resulting from water injection across different oil layers were not taken into account. As a result, the major accident hazard of ultra-high pressure emerged in multiple oil layers.

  (d) The excess water was injected into the well B23 to sustain the pressure level. As a result, the pre-existing geological fault 502 was activated, resulting in its extension to the seafloor associated with oil spills.

- *Escalation prevention of the environmental damage failed.* The process of identifying potential sources of oil spills and sealing leaks progressed at a sluggish pace, extending over a period of more than three months.

The following barrier failures have been identified as contributing factors to the C-Accident:

- *The well integrity either failed or was not established.*

  (a) The ODP regulations regarding the debris reinjection were not adhered to. Specifically, the debris reinjection layer was adjusted upward several times

Table 3.2 The barrier performance in the Chinese Penglai 19-3 oil spill accidents

Barrier function no.	Barrier function	Performance
BF.1	Maintaining the integrity of the well system	Failed
BF.2	Preventing ignition	Successful
BF.3	Reducing cloud and spill size	Failed
BF.4	Preventing escalation	Successful
BF.5	Preventing fatalities	Successful

towards the vicinity of the reservoir without proper authorization. This led to an ultra-high-pressure hazard in the proximity of the reservoir's base.

(b) The setting depth of the surface casing for the well C20 was not designed in accordance with the design requirements outlined in the environmental impact assessment (EIA) report.

(c) The pressure-bearing capacity of the surface casing shoe was inadequate, resulting in lateral leakage. This further led to the formation fracture with oil spills on the seabed during drilling of the well C20 into the over-pressured layer.

- *Escalation prevention of environmental damage failed.* Intermediate casings were not installed during the drilling of well C20 into the over-pressured formation. The process of identifying potential sources of oil spills and sealing leaks progressed at a sluggish pace, extending over a period of more than 6 months.

Table 3.2 provides a summary of the barrier performance for the Chinese PL 19-3 oil spill accidents.

Lessons Learned

The lessons learned from the PL 19-3 oil spills in China emphasize two key areas: well design and well operations.

The lessons pertaining to well design are summarized in the following

- Pressure-measuring tests were not conducted to verify the stability of pre-existing geological faults.
- The upper safety limit for the pressure at the water injection wellhead was not established.
- The surface casing for the well C20 was not installed to the setting depth in accordance with the EIA requirements.

The lessons pertaining to well operations are summarized in the following.

- Ineffective leadership,
- Fragmentation of information coupled with insufficient communication,
- Deficiencies in the management of changes (MOC) controls,

- Lack of compliance,
- Ineffective oversight and management of contractors,
- Failure to provide timely procedures,
- Inadequate training and supervision of staff,
- Inability to properly analyze and evaluate risks,
- Unidentified deficiencies in design,
- Emphasis on time and costs over the management of major accident risks.

The SOA investigation report clearly pointed to a primary failure in leadership and communication within the organization. The COPC's offshore team breached the ODP rules without reporting this violation to management. Especially, the drilling team of the well C20 was not informed that the debris reinjection layer in the well C25 was adjusted upward to the vicinity of the reservoir without authorization. As a result, the well C20 was drilled into a fault block characterized by ultra-high pressure, resulting in a well kick incident. These failures highlight the deficiencies in MOC control and the lack of compliance as well. Furthermore, the SOA investigation report revealed the ineffective oversight and management, noting that the COPC did not comply with the ODP and EIA criteria. Additionally, there was a lack of effective oversight and management from the regulatory authority or CNOOC to address these deficiencies. The SOA investigation report revealed a number of managerial deficiencies directly or indirectly, including a lack of adequate emergency plans, failure to provide timely procedures, as well as inadequate training and supervision of staff. The COPC did not develop formal emergency plans to interpret and deal with various abnormal situations. For instance, during the B-Event, there was a significant increase in the water-injection rate for well B23, accompanied by a dramatic drop in the injection pressure. This suggests that staff had not received training on interpreting and dealing with such an abnormal situation. This also indicated a broader lack of risk awareness. The important early warning signals were overlooked by the operating staff. Although not explicitly stated in the SOA report, it can be inferred that deficient work practices were prevalent within COPC. Many decisions and actions appeared to prioritize time and cost factors over control of major accident risk. This may explain the weakness in design and MOC control in ensuring the well integrity.

In examining the important lessons learned pertaining to emergency responses, it becomes evident that the major accident hazard of geological oil spills had not been identified within the Chinese offshore petroleum sector. Actually, the term "geological oil spill" was not formally defined until the occurrence of the PL 19-3 oil spills (Yue et al. 2018). Consequently, it is reasonable that emergency response plans were not developed in response to geological oil spills. Furthermore, as illustrated in the timeline presented in Fig. 3.2, production activities at Wellhead Platforms B and C were not suspended until one month later, while the injection, drilling and production activities in the PL 19-3 oilfield persisted for an additional 3 months prior to the final shutdown. This delay underscores the inadequacy in determining critical parameters of emergency responses, specifically the response time and the necessary capacity for emergency actions.

3.2.2 US DWH Oil Spill

Background

In March 2008, BP Exploration & Production Inc. (commonly referred to as BP) acquired the lease for the Macondo well, located in the central GoM, approximately 50 miles from the nearest coastline in Louisiana. The water depth at the site of the well is about 1500 m. As the operator, BP held the responsibility for the design and development of the well, which was intended to reach a total depth of approximately 6157 m.

In October 2009, BP commenced the drilling activities utilizing the Transocean Marianas rig. However, on November 9, 2009, the rig sustained substantial damage due to Hurricane Ida, necessitating its relocation to a shipyard for repairs. In this situation, the Transocean DWH rig was deployed as a replacement for the Marianas rig.

The DWH rig was an advanced semi-submersible Mobile Offshore Drilling Unit (MODU) with the capacity to accommodate 146 crew members. It entered service in 2001. The DWH rig was designed with a maximum operational capacity of 2438 m of water depth and 9144 m of drilling depth. The rig, predominantly used by BP, successfully drilled more than 30 wells on the U.S. outer continental shelf (OCS). Notably, the DWH rig maintained an excellent record of preventing lost-time accidents. BP afforded an operational cost of approximately 1 million dollars per day for the DWH rig.

The DWH rig arrived at the drilling site on 31 January 2010, and drilling operations commenced on 11 February 2010. On 20 April 2010, the Macondo drilling activities were completed after several productive oil and gas zones had been discovered, and the crew members were carrying out the procedure to temporarily abandon the well. The well control was lost in the evening, resulting in the influx of hydrocarbons into the drilling riser, which subsequently reached the DWH rig. Later, an explosion and subsequent fires occurred, leading to the tragic loss of eleven crew members and critical injuries to 17 others. The rig continued to burn for approximately 36 h before ultimately sinking on 22 April 2010. It is estimated that five million barrels of oil leaked from the damaged well. On 4 April 2016, a federal court in New Orleans sanctioned a landmark settlement amounting to $18.7 billion, thereby resolving all legal disputes between the federal government and the five Gulf States regarding their economic and environmental claims related to the DWH oil spill. BP allocated over $69 billion in relation to the oil spill, encompassing expenses for response efforts, cleanup operations, economic claims, government payments, settlements and restoration initiatives (BP 2022).

Sequence of Events

Figure 3.3 illustrates the timeline of the DWH oil spill accident, encompassing the period from the initial blowout to the ultimate sealing of the well, which lasted for 87 days.

On 20 April 2010, at approximately 8:00 p.m., the blowout preventer (BOP) was activated to open, and seawater pumping started through the drill pipe to displace mud

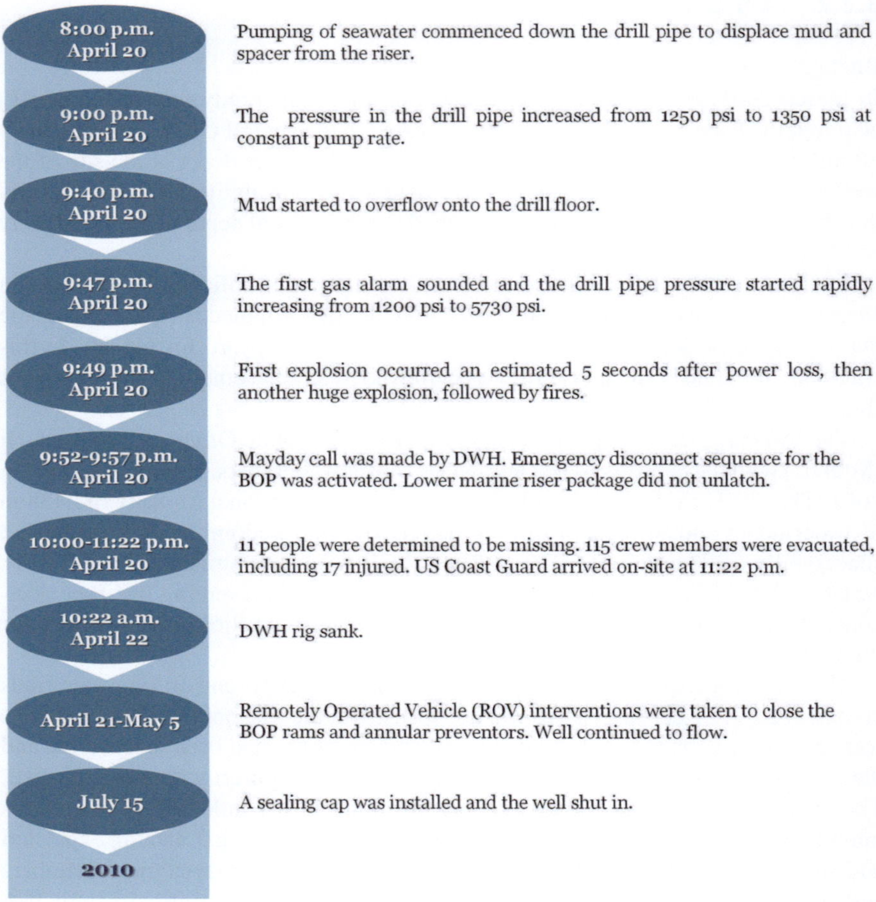

Fig. 3.3 Timeline for the DWH oil spill accident

and spacer from the riser. This operation resulted in a reduction of pressure within the drill pipe. However, soon after 9:00 p.m., an abnormal increase in drill pipe pressure was observed despite the constant pump rate. This increase was unexpected, as the pressure was anticipated to decrease due to the replacement of the denser mud with the less dense seawater. Over the subsequent 40 min, there were several clear signals, suggesting the influx of hydrocarbons into the well. Despite these warning signs, the crew members seemed unaware of the hazardous situation and no apparent control actions were taken.

At 9:40 p.m., the crew members were alerted to the occurrence of a kick when an overflow of drilling mud was observed on the drill floor. In reaction to this urgent incident, the crew members took immediate measures. The flow was diverted to the mud gas separator (MGS). However, the MGS was not designed to handle the well flow and quickly became overwhelmed. Concurrently, the annular preventer on the BOP was activated in an effort to shut in the well. Unfortunately, these interventions were too late to prevent an explosion, as gas had flooded onto the rig by that time. The first gas alarm was triggered at approximately 9:47 p.m. Gas rapidly dispersed and activated additional alarms. At approximately 9:49 p.m., the initial explosion occurred within five seconds following a loss of power. This was succeeded by a second explosion occurring within ten seconds of the first explosion. Subsequently, a fire ensued. At 9:52:57 p.m., the DWH rig issued a mayday call. The emergency disconnect system (EDS) was not activated successfully, and the lower marine riser package (LMRP) could not be unlatched. Between 10:00 p.m. and 11:22 p.m., eleven crew members were reported missing, and search and rescue activities were carried out. The remaining 115 crew members were evacuated, among whom 17 sustained injuries. The rig was engulfed in flames for approximately 36 h before it sank at 10:22 a.m. on 22 April 2010.

Between 21 April and 5 May, many attempts to close the BOP rams and annular preventers by use of the remotely operated vehicles (ROVs) were unsuccessful, resulting in the well's continued flow. On 15 July 2010, a sealing cap was successfully installed, finally shutting in the well. At that time, the oil spill from the Macondo well persisted for a duration of 87 days.

Barrier Performance in the Accidents

The BP investigation report (BP 2010) delineates the failures of barriers, as depicted in Fig. 3.4, and specifically addresses the following aspects.

- *The well integrity either failed or was not established.* The failure of this barrier function mainly involves the failure of two barrier elements.

  (a) The hydrocarbons were not isolated successfully by the annulus cement barrier.
  (b) The hydrocarbons were not isolated successfully by the shoe track barriers.

- *Hydrocarbons entered the well undetected and well control was lost.* The failure of this barrier function mainly involves the failure of three barrier elements.

  (a) The negative-pressure test (NPT) was conducted even though well integrity had not been established.
  (b) The influx of hydrocarbons was not detected until they entered the riser.
  (c) The measures taken to regain control of the well were unsuccessful.

- *Hydrocarbons ignited on the DWH rig.* The failure of this barrier function mainly involves the failure of two barrier elements.

  (a) The diversion of gas to the MGS led to the release of gas onto the rig.
  (b) The fire and gas detection system failed to prevent the ignition of hydrocarbons.

- *BOP failed to seal the well prior to and after the accident*. Three approaches employed to activate the BOP in the emergency mode were proven ineffective, encompassing crew-initiated EDS function, crew-initiated automatic mode function (AMF) and ROV-initiated auto-shear function.

Table 3.3 provides a summary of the barrier performance for the US DWH oil spill accident.

Lessons Learned

The lessons learned outlined below are a concise summary of the SINTEF report (Tinmannsvik et al. 2011).

An obvious question on the DWH accident arises: *how was it possible for so many safety barriers to fail at the same time*? The answers are not straightforward

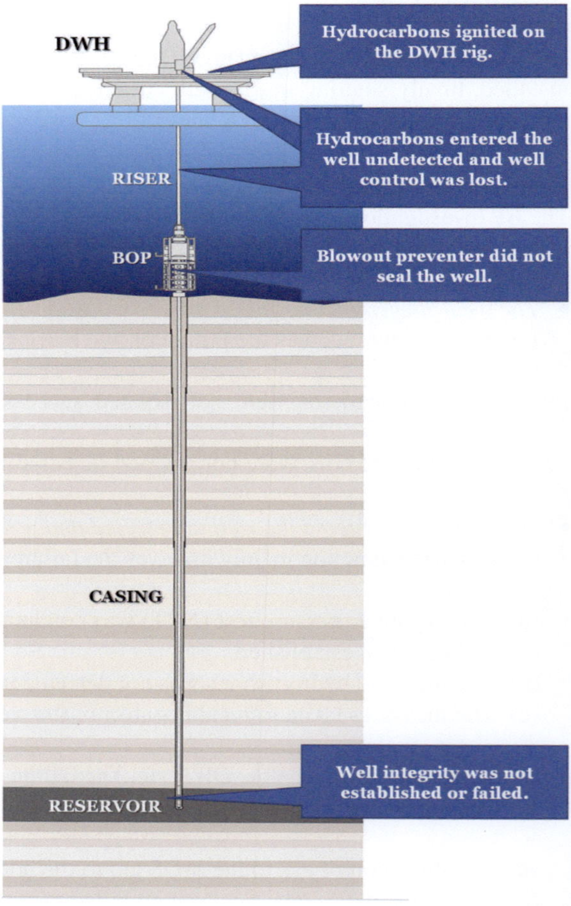

Fig. 3.4 Barrier failures leading to the DWH oil spill accident. *Source* Adapted from OLF (2012)

Table 3.3 The barrier performance in the US DWH oil spill accident

Barrier function no.	Barrier function	Performance
BF.1	Maintaining the integrity of the well system	Failed
BF.2	Preventing ignition	Failed
BF.3	Reducing cloud and spill size	Failed
BF.4	Preventing escalation	Failed
BF.5	Preventing fatalities	Failed

but correlate to two key factors: maintaining control of the barriers and managing an increasing degree of complexity.

- The management of barriers throughout the lifecycle of a well presents significant challenges, primarily due to the dynamic nature of the safety issues and the large number of different well operations.
- An increasing degree of complexity is identified as a common characteristic pertaining to the failure of barriers. This finding can be exemplified by the need for numerous involved actors to interact, frequent reorganizations, the introduction of new work processes and rapid technological advancements. Consequently, it is not surprising that offshore drilling operations in complex reservoirs are often characterized by an ongoing procedure pertaining to problem-solving, wherein novel and unforeseen challenges must be addressed in real-time. The escalating complexity imposes additional requirements on our conceptualization of safety, raising concerns about whether the safety research has adequately adapted to the rapid pace of technological innovation. Therefore, it is imperative to enhance the capabilities for interaction and decision-making within an increasingly complex and uncertain environment.

Furthermore, the subsequent organizational factors highlight significant lessons on enhancing safety.

- *Ineffective leadership.* BP management did not implement adequate safety oversight and failed to consistently adhere to the established safety regulations and procedures.
- *Compartmentalization of information and deficient communication.* The onshore team at BP and the drilling supervisor did not communicate with one another concerning the risks associated with cementing.
- *Failure to provide timely procedures.* Neither BP nor Transocean had established internal and formal procedures for the execution and interpretation of the NPT results.
- *Poor training and supervision of employees.* BP and Transocean did not offer any official training regarding the appropriate procedures for conducting and interpreting an NPT.

- *Ineffective management and oversight of contractors.* Even though BP was aware of the uncertainties surrounding Halliburton's cement testing procedures, it failed to adequately address these deficiencies.
- *Inadequate use of technology/instrumentation.* The well monitoring apparatus on the DWH rig was inadequate, and neither BP nor the other enterprises involved made adequate use of the information provided by the available data displays and monitoring devices. This inadequacy significantly contributed to the inability of the crew members to promptly detect the kick.
- *Inadequate analysis and perception of risk.* The management system at BP required separate risk analyses to be performed during the planning phase of the well; however, such analyses were not required during the execution phase. Consequently, significant decisions were made during the execution phase without the evaluations from formal risk assessments. Concurrently, the crew members operated under the assumption that they possessed comprehensive awareness of all potential hazards, despite the fact that they were unable to maintain adequate oversight of these hazards, indicating a deficient risk perception.
- *Improper MOC.* Modifications preceding implementation of drilling procedures generally do not undergo rigorous peer review or an MOC process.
- *Prioritizing time and costs over management of major accident risks.* BP prioritized time and cost savings over safety in a lot of its operational decisions. At the time of the blowout, the operation had experienced a delay of 38 days and exceeded the budget by approximately $58 million. This context may elucidate the insufficient emphasis on ensuring the well integrity. On the day of the accident, the team on the DWH rig was commended by BP management for achieving seven years of drilling without any "lost time incidents". However, the investigation into the Texas City explosion in 2005 indicated that BP demonstrated a significant commitment to personal safety, but less focus on major accident risks.

3.3 Lessons from Piper Alpha Explosion and Fire and FPSO Explosion

This section aims to present and discuss the critical implications for the prevention and mitigation of offshore process accidents in accordance with two notable accidents in the UK and Brazil, respectively: the explosion and fire on the Piper Alpha platform and the explosion on the FPSO unit CDSM.

3.3.1 UK Piper Alpha Explosion and Fire

Background

The Piper Alpha platform, a large fixed installation, was located in the Piper field, approximately 193 km northeast of Aberdeen, Scotland. It operated at a depth of 144 m.

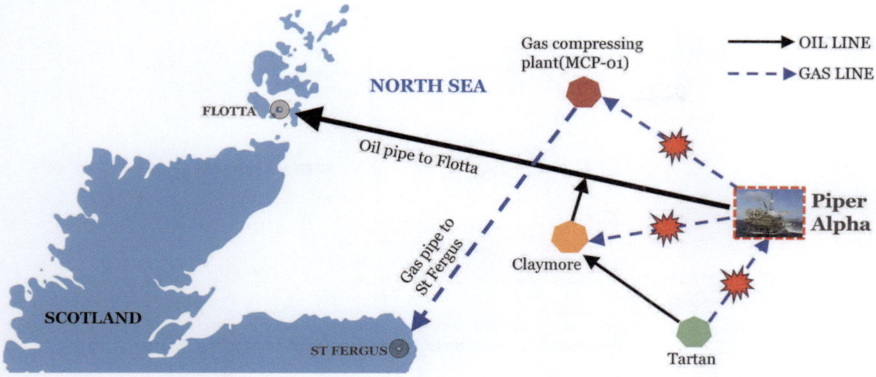

Fig. 3.5 The pipeline connections to the Piper Alpha platform

The Piper Alpha platform commenced its production activities in late 1976, initially focusing solely on oil production while flaring the associated gas. In 1980, the Piper Alpha platform underwent retrofitting, enabling it to facilitate the production of both oil and gas.

Figure 3.5 illustrates the pipeline connections associated with the Piper Alpha platform, which was primarily equipped with four transport risers, namely the oil export riser, the Claymore gas riser, the Tartan gas riser, as well as the MCP-01 gas riser. It is noteworthy that none of the risers had riser emergency shutdown valves (ESDVs) so as to protect the Piper Alpha platform from the large amount of oil and gas within the pipelines.

The topsides configuration of the Piper Alpha platform is shown in Fig. 3.6. The Piper Alpha platform consisted of the following four modules. Notably, the design of the platform incorporated firewalls between the A, B, C, and D modules; however, the blast walls were absent.

- Module A: the wellhead module, which consisted of 36 wellheads and was organized into three rows, each containing 12 wellheads.
- Module B: the separation module, which consisted of 2 main production separators.
- Module C: the gas compression module, which consisted of two high-pressure condensate injection pumps (A and B). Typically, one pump was actively operational while the other remained in a standby mode. There was no automated system for switching between the pumps. In the event of a pump trip or failure, manual activation of the standby pump was necessary. Pressure safety valves (PSVs) 504 and 505 were installed on Pump A and Pump B, respectively.
- Module D: the power generation module, which consisted of the main electrical generators.

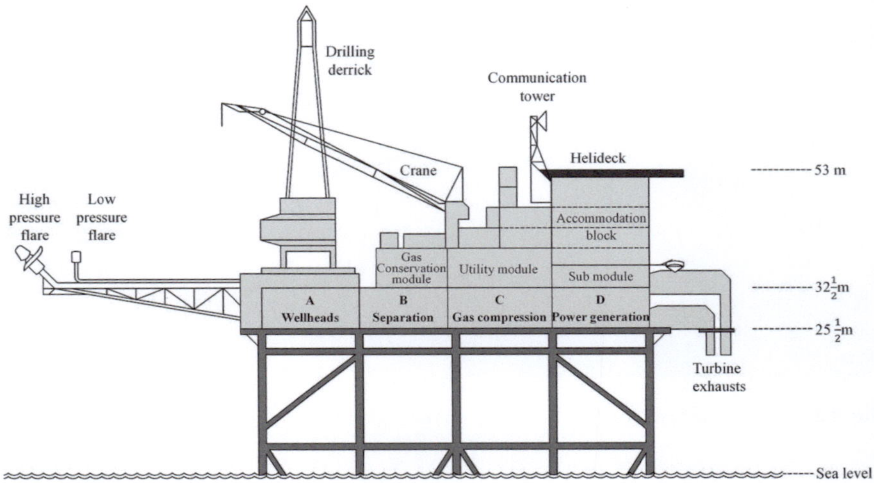

Fig. 3.6 The topsides layout of the Piper Alpha platform

On the evening of 6 July 1988, an explosion followed by a fire on the Piper Alpha platform resulted in the fatalities of 165 out of the 226 individuals present, as well as 2 crew members aboard one of the fast rescue crafts (FRCs). As of now, the number of fatalities remains the highest ever recorded in the global offshore petroleum industry.

Sequence of Events

Figure 3.7 illustrates the timeline of the Piper Alpha accident, covering the period from the communication failure during the shift handover to the ultimate structural collapse of the platform, which lasted a total of 6 h and 45 min. The subsequent section presents a concise summary of the principal events highlighted in the public inquiry into the Piper Alpha disaster (Cullen 1990).

On the morning of 6 July 1988, the PSV 504 associated with the condensate pump A was taken offline for maintenance under a separate permit. Subsequently, a blind flange, which was only hand-tightened, was applied to seal the exposed condensate pipe temporarily. The engineer on duty made the note on the work permit that the condensate pump A was out of service, indicating that the maintenance activity could not be completed before the scheduled end of the day shift at 6:00 p.m. However, this information was not effectively communicated to the night shift personnel.

At approximately 9:45 p.m. on the same day, the condensate pump B ceased operation and could not able to be rebooted, likely due to a blockage caused by methane clathrate. The night shift did not realize that the condensate pump A was unsafe and decided to start it. Subsequently, at around 9:55 p.m., the condensate pump A was activated, resulting in gas entering the pump and escaping from a blind flange under high pressure. This incident triggered multiple gas alarms. Before any corrective measures could be implemented, the gas ignited, leading to an explosion. The firewall was damaged by the first explosion within Module C, and a weak condensate

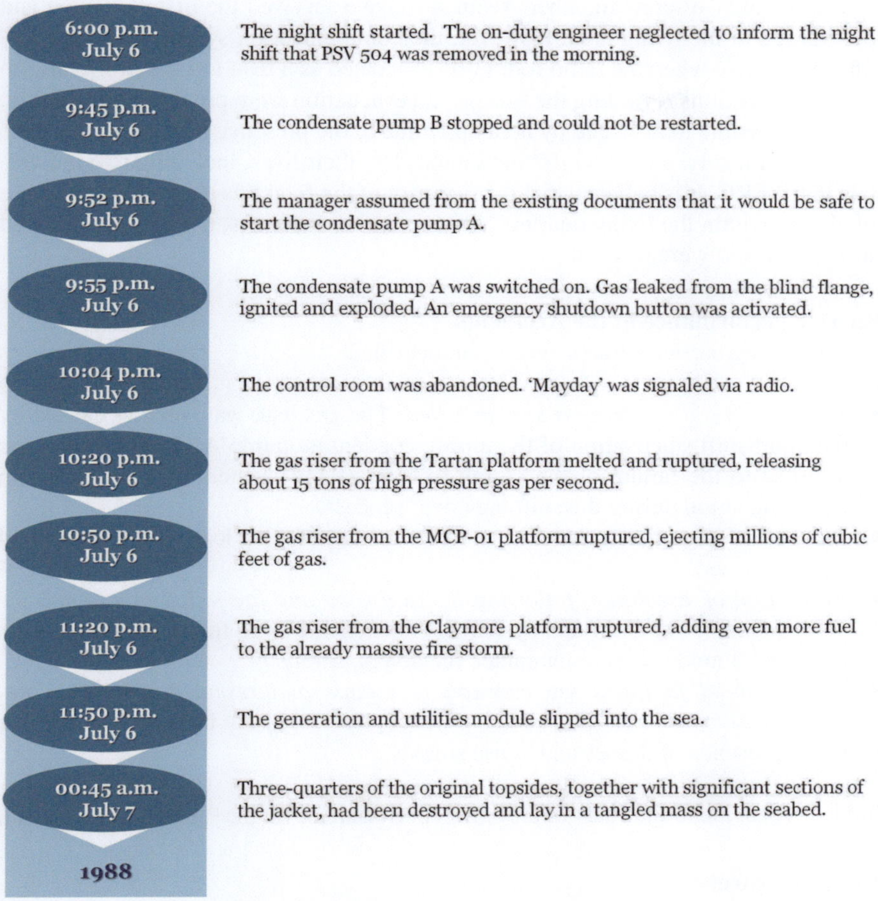

Fig. 3.7 Accident timeline for the Piper Alpha platform

pipe was ruptured in the adjacent Module B, which in turn ignited secondary fires. To safeguard divers, all fire pump switches had been set from the automatic mode to the manual mode, rendering the fire water system ineffective and exacerbating the situation. At about 10:20 p.m., a significant explosion was triggered by the rupture of the Tartan gas riser, resulting in a sustained and intense high-pressure gas fire. A further substantial explosion transpired at around 10:50 p.m. as a result of the rupture of the MCP-01 gas riser, intensifying the existing fire. By 11:20 p.m., the rupture of the Claymore gas riser contributed additional fuel to the already severe conflagration. At this juncture, the structural integrity of the Piper Alpha platform had deteriorated to such an extent that it started to collapse. By approximately 12:45 a.m. on 7 July 1988, three-quarters of the original topsides, along with significant portions of the jacket, had been obliterated.

The majority of crew members seemed to have survived the first explosion and took refuge in the living area. However, the communication system sustained significant damage when the radio room was evacuated as a result of the fire. Thus, no further instructions regarding the escape and evacuation were provided. In addition, it was impossible to evacuate by helicopter due to the presence of flames and dense smoke. Ultimately, a total of 167 individuals lost their lives, including two individuals in the FRC. It is believed that the majority of the fatalities resulted from smoke inhalation within the living quarters. Among the 61 survivors, the majority jumped into the sea and were rescued.

Barrier Performance in the Accidents

The following barrier failures have been identified:

- *The integrity of the process system failed.* The gas leak was directly caused by the night shift's activation of the unsafe condensate pump A. This incident was ascribed to the failure of the permit-to-work (PTW) system and the failure in communication during the shift handover process.
- *Prevention of ignition failed.* The source of ignition for the leak was not identified and confirmed.
- *Prevention of escalation following the explosion and fire failed to mitigate the consequences.* The fire water system did not activate as all fire pumps were set to the manual mode so as to guarantee the divers' safety.
- *Prevention of fatalities* via *evacuation, escape and rescue failed.* The pre-established evacuation plan utilizing helicopters could not be implemented due to the presence of flames and dense smoke.

Table 3.4 provides a summary of the barrier performance for the Piper Alpha accident.

Lessons Learned

The lessons learned outlined below are a concise summary derived from the well-known Cullen Report (Cullen 1990).

- *Design issues.*

Table 3.4 The barrier performance in the Piper Alpha accident

Barrier function no.	Barrier function	Performance
BF.1	Maintaining the integrity of the process system	Failed
BF.2	Preventing ignition	Failed
BF.3	Reducing cloud and spill size	Failed
BF.4	Preventing escalation	Failed
BF.5	Preventing fatalities	Failed

(a) The central control room and the living area were designed to be situated in proximity to the gas compression module, and lacked adequate protection from the fire and blast barriers.
(b) The unprotected pipes were designed and thus lacked the capacity to endure prolonged exposure to fire loads.
(c) The living area was situated at an elevated position, which facilitated the rapid accumulation of smoke.
(d) The platform underwent modifications to enable the processing and exportation of gas, thereby introducing an additional risk of explosion. However, the firewalls were not upgraded to blast-proof walls.

- *Major accident risk* versus *occupational accident risk*. To ensure the safety of divers, the fire water system did not activate, as all fire pumps were switched from the automatic mode to the manual mode. The occupational accident risk posed by a close diver sucked into a pipe was prioritized over the major accident risk associated with leaving 226 crew members exposed to fire.
- *Compartmentalization of information and deficient communication*. The PTW system depended on informal communication, and the communication between shifts was deficient. Specifically, the night shift was not informed that the PSV 504 had been taken out for overhaul and had not yet been replaced.
- *Emergency response*. A variable of escape routes to the sea should be provided on the installation. Many personnel who jumped into the sea were successfully rescued.

3.3.2 Brazilian FPSO Unit CDSM Explosion

Background

The FPSO unit CDSM was situated at a depth of 792 m within the Espírito Santo Basin, facilitating petroleum production from the Camarupim and Camarupim Norte fields. These fields are estimated to hold around 4.4 billion cubic meters of gas and 0.7 million cubic meters of condensate. The unit underwent conversion from the oil tanker NAVARIN during the period from 2007 to 2008. BW Offshore operated this FPSO unit, which was chartered by Petrobras after its conversion.

The FPSO unit CDSM primarily focused on the production of natural gas, utilizing connections from four gas wells. Within the gas processing system, the gas underwent a series of separation processes through various separator vessels, which effectively separated the gas from the condensate. The processed gas was transported via pipeline to the onshore Gas Treatment Unit. Occasionally, the condensate was exported through the pipeline. Typically, deck cargo tanks store the condensate for the purpose of exportation by means of tanker transport.

The FPSO unit CDSM comprised seven cargo tanks for the storage of produced fluids. Among these seven cargo tanks, five (designated as 2C-6C) were primarily utilized for the storage of liquid products intended for subsequent export to tankers. The remaining two cargo tanks were designated for the temporary storage of dirty and clean water, respectively, referred to as slop dirty and slop clean tanks. The pump room was situated adjacent to both the engine room and the storage tanks.

On 11 February 2015, an explosion took place in the pump room, resulting in the fatalities of 9 out of 74 crewmen on board. In addition, 26 crewmen were injured, and the FPSO unit suffered significant damage.

The investigation report conducted by ANP (2016) is the primary source for the following discussions.

Sequence of Events

Figure 3.8 illustrates the timeline of the FPSO unit CDSM accident, outlining the sequence of events from the initial condensate leak to the explosion, all of which occurred within a one-hour period.

On the morning of 11 February 2015, a mixture of water and condensate from the central cargo tank 6C was transferred to the port waste tank by use of the stripping pump located in the pump room. This operation commenced at approximately 8:53 a.m.

At approximately 11:30 a.m., a condensate leak in the pump room was captured by the closed-circuit television (CCTV) camera 5. However, this leak went unnoticed by operators in the control room. Upon investigation, it was determined that the leak originated from a failure of the spade within the flange connection of the piping system.

The gas was promptly detected by three gas detectors installed at the base of the pump room. At 11:35:37 a.m., one detector indicated that the Lower Explosive Limit (LEL) had reached 100%. Nevertheless, the production facility continued its operations, as halting production required a manual intervention based on an evaluation from the emergency command.

At 11:44 a.m., the alarms were deactivated, and three gas detectors were temporarily disabled to minimize auditory distractions during emergency communications. Subsequently, at 11:49 a.m., the first response team was dispatched to the pump room to ascertain the origin of the gas leak. Upon their descent, the firefighters' portable detectors were triggered, warning the combustible gas. It was identified that the leak originated from a flange situated upstream of a valve linked to a slop tank. The first team observed a condensate pool covering an area of approximately 2 m^2 on the ground. By 11:54 a.m., the first team had made its way back to the main deck.

At 12:07 p.m., the second team re-entered the pump room to evaluate the necessary services for repairing and restoring normal operations. During their descent, a portable detector indicated a 100% LEL. Approximately eight minutes later, the second team made an exit, believing that this hazardous event was controllable. Consequently, individuals at the muster points were released from their assembly.

At 12:27 p.m., the third team arrived in the pump room, carrying equipment including absorbent blankets, a firefighting hose, and a ladder. One minute following the descent, a portable detector indicated that the LEL had reached 100%. The absorbent blankets and firefighting hose were employed for cleaning the site, while a ladder and tools were used to fix the leaky flange connection. The initial plan of utilizing absorbent blankets to soak up the leaked condensate did not work and

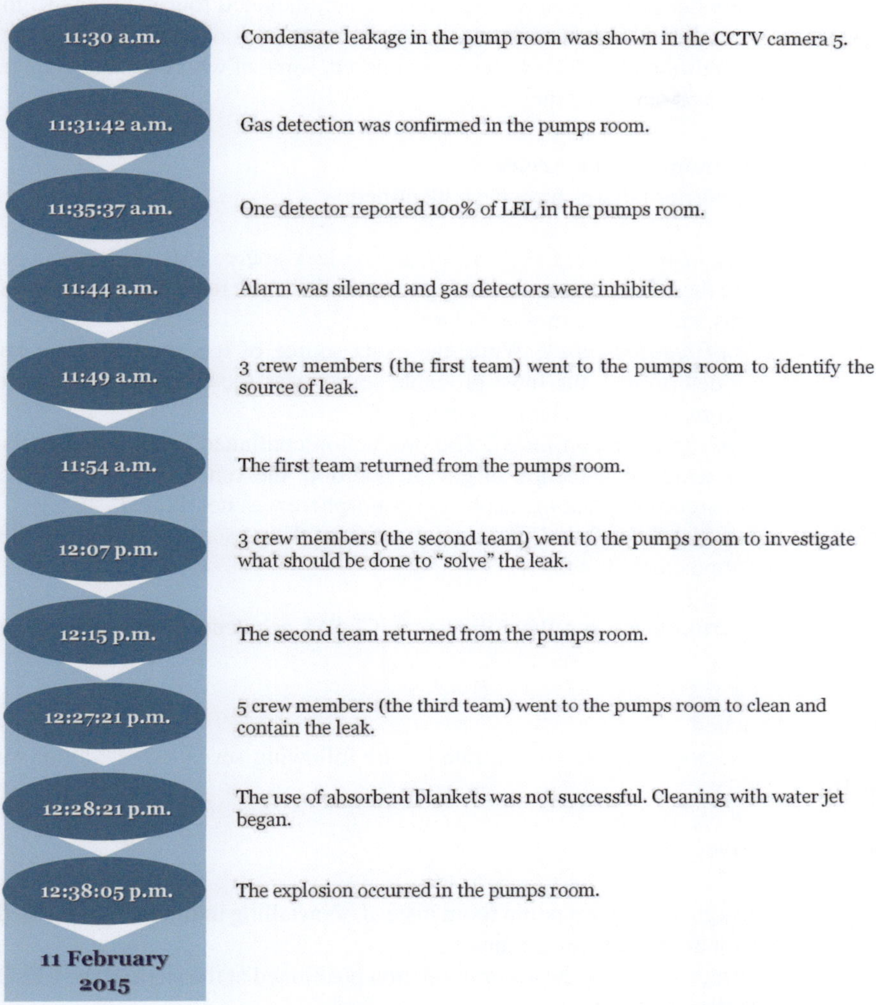

Fig. 3.8 Timeline for the FPSO CDSM accident

was discontinued. Consequently, the third team resorted to employing a firefighting hose to wash down the condensate into the drainage system. A new stretch of 15 m was attached to the firefighting hose, and a request was made to increase the water pressure.

At approximately 12:38:05 p.m., a powerful explosion took place. The shock wave caused the bulkhead to rupture, leading to significant harm to the pump room, engine room, access shaft, main deck surrounding the access shaft, and the living quarters. The most probable source of ignition was identified as the electrostatic current generated from the nonconductive water jet.

The total number of fatalities reached nine, which included four members from the third team, four individuals on the main deck, as well as one individual in the engine room. Additionally, 26 crewmen were injured, some of whom sustained their injuries within the accommodation area.

Barrier Performance in the Accidents

The following barrier failures have been identified:

- *The integrity of the process system failed.* The leak source was identified as a flange due to the failure of the installed spade, which did not meet the specification requirements and the pipe-pressure class.
- *Prevention of ignition failed.* While the exact source of ignition could not be definitively determined, the most probable source was the electrostatic current generated from the nonconductive water jet.
- *Reduction of the gas cloud failed.* The production continued for approximately 10 min following the detection of gas. In addition, the ventilation system was deactivated, signifying that the explosive atmosphere was not diluted.
- *Prevention of fatalities failed.* The minimization of the personnel exposure to the explosive atmosphere failed.

The barrier performance for the FPSO unit CDSM accident is summarized in Table 3.5.

Lessons Learned

The primary lessons learned concentrate on the following seven aspects (Vinnem 2018).

- *Design issues.*

  (a) It is designed to have a pump room instead of installing individual submerged pumps within each storage tank.
  (b) It is designed to have the accommodation positioned at the stern of the FPSO unit instead of the bow.

Table 3.5 The barrier performance in the FPSO unit CDSM accident

Barrier function no.	Barrier function	Performance
BF.1	Maintaining the integrity of the process system	Failed
BF.2	Preventing ignition	Failed
BF.3	Reducing cloud and spill size	Failed
BF.4	Preventing escalation	Successful
BF.5	Preventing fatalities	Failed

- *Inadequate storage of the condensate.* The cargo tanks were designed for the storage of oil or a mixture of oil and condensate. Nonetheless, these tanks were used to store condensate without conducting any hazards or risk review pertaining to the potential impact of operational modifications.
- *Degradation of the mechanical system.* The cargo tank system is expected to experience degradation over time when storing various fluids, in comparison to its original design specifications. Notably, the failure of the tank valve seal was attributed to chemical destruction caused by the aromatic hydrocarbons present in the condensate.
- *Degradation of the marine staff*. The marine staff was also being degraded over time as the operational constraints resulting from the operation of a degraded system are necessary.
- *Manipulating the stripping pump while the associated discharge valve was closed.* The operator manipulated the stripping pump indirectly by monitoring the vapor control valve's opening percentage and the discharge pressure, rather than focusing on the number of strokes. The opening percentage of the vapor control valve influenced the pump discharge pressure, which was monitored by the operator to know whether the discharge valve was opened. However, the correlation between the vapor control valve's opening percentage and the discharge pressure could be confused, leading to potential misdiagnosis of high pressure in the pump discharge.
- *Personnel Exposure.* Three teams were exposed to the explosive environment. Notably, following the return of the second team, the emergency management director made a decision regarding the demobilization of meeting points. This decision significantly increased personnel exposure.
- *Ignition of the flammable atmosphere.* The possible ignition sources induced by the emergency response actions were not taken into account at all.

3.4 Importance of Root Causes

There is no universally accepted definition of root cause, as different researchers employ differing levels of causation in their analyses. Up to the present, the most relevant definition for root cause is the most basic cause that can be reasonably identified, and that management has control over to fix. This definition contains three key elements (Paradies and Busch 1988; HSE 2001):

- Basic Cause. This refers to specific reasons as to why an accident occurred that enable recommendations to be made, which will prevent recurrence of the events leading up to the accident.
- Reasonably Identified. Accident investigation must be completed in a reasonable time frame.
- Control to fix. The identified causes should be specific enough to allow those in charge to rectify the issue.

It should be noted that basic cause is a broader and less precise term than root cause, and is widely used in various areas. Notably, root causes are the most essential outcome derived from accident investigation. Thus, in accident investigations, the focus should be on the term "root cause" rather than "basic cause". Moreover, it has been confirmed that many accidents share a significant number of common root causes. Thus, identification of root causes is of great importance to help management effectively learn lessons from past accidents, thereby facilitating the implementation of corrective actions and minimizing the likelihood of an analogous accident occurring in the future.

Typically, an accident investigation report concentrates on the direct causes and underlying causes, while root causes are missing. The direct cause and underlying cause represent the most and less obvious reasons, respectively, for an accident occurring. The following is a case of the direct cause and underlying cause from the investigation report on the hydrocarbon leak at the Oseberg A platform (Havtil 2013).

- Direct cause: Sand erosion led to a hole.
- Underlying cause: The adequate review of the plant had not yet been performed to verify its capability to handle sand production.

Some researchers or readers may wonder whether this underlying cause is indeed the root cause. The answer is no, as it does not represent the basic cause that management has control over to fix. Possible root causes could be "focus on time and costs", or "lack of supervision".

In accident investigation reports, the differentiation between root causes and barrier failures is often ambiguous. This is primarily because barriers are defined rather loosely, while the more precise classification of concepts regarding the barrier function and barrier element is usually not preferred. The concept of barriers will be discussed in greater detail in the Seventh Lesson of this book.

3.5 Accident Investigation Process

The primary objectives of investigating accidents, as outlined by Leveson (2011), are twofold:

(1) To assign blame for the accident;
(2) To understand why it happened so that analogous accidents can be prevented.

As has been stated in Sect. 3.1, this book concentrates on the second objective, because this is the way to improve safety. In order to achieve this objective, an accident investigation process encompasses a comprehensive array of activities, which are described in different ways by various researchers or parties. The subsequent discussions primarily draw upon the US Department of Energy (DOE)

Workbook (U.S. DOE 1999), which offers comprehensive instructions for carrying out accident investigations.

The DOE investigation process commences promptly following the event's occurrence and concludes with the endorsement of the final investigation report by the designated authority. Overall, the DOE investigation process is divided into three principal phases:

- Collection of evidence and facts;
- Analysis of evidence and facts;
- Development of judgments of need and reporting.

It can be noted that the three principal phases within an accident investigation exhibit overlap, and the entire process is characterized by its iterative nature. This section will focus on and explore these three principal phases of activities involved in an accident investigation.

(1) **Collection of evidence and facts**.

The initial phase involves collecting data, serving as the basis for the whole process of accident investigation. It is essential to collect three primary categories of evidence:

- Human evidence, which includes testimonies and observations provided by witnesses;
- Physical evidence, which refers to physical items like debris, equipment, etc.
- Documentary evidence, which comprises both electronic and paper records.

Notably, the process of collecting data can often be lengthy, time-consuming and piecemeal. Human evidence may be incomplete or contradictory pertaining to the detailed description of the accident. Additionally, physical evidence could be severely compromised or completely destroyed, and documentary evidence could be limited or difficult to acquire. Against this challenge, a thorough investigation is required to achieve a sufficiently comprehensive understanding of the accident.

Moreover, it is imperative that the accident investigation conducts a comprehensive examination of the organizational issues, management framework, and the supervisory procedure employed by line management. This examination seeks to determine whether deficiencies in these areas played a role in the causes of the accident.

Lastly, evidence preservation and control are essential for ensuring the evidence's accuracy and validity, as well as for upholding the integrity and credibility of the investigation.

(2) **Analysis of evidence and facts**.

In this phase, the collected data should be comprehensively analyzed to identify the causal factors. The answers to the following three questions should be sought:

- What has occurred?
- What is the reason for its occurrence?
- How is the way it occur?

(3) **Development of judgments of need and reporting**.

Judgments of need refer to the management controls and safety measures that are needed to reduce the likelihood of an analogous accident occurring in the future. Each judgment of need ought to be associated with causal factors, analyses, and evidence. In instances where this association is weak, the corresponding judgement of need should be disregarded.

The investigation report aims to present the investigation results in a clear and concise manner, thereby facilitating the reader's comprehension of the events that transpired, the underlying reasons why they occurred, and the measures that can be implemented to prevent analogous accidents in the future. The investigation report is presented without assigning individual blame or recommending punitive actions. The following two points from Hendrick and Benner (1987) are worth taking into account concerning the investigation report.

- Investigations are remembered through the reports;
- The best investigation will be wasted by a poor report.

3.6 Enterprise Investigations: Internal or External (Independent)?

In some countries, such as the U.S. and Norway, both internal enterprise investigations and external (independent) investigations may be conducted and made publicly available.

When these two types of investigations are compared, it is often found that independent investigations obviously will be wide-ranging in considering potential causes, and possibly considering more extensively potential outcomes in slightly different circumstances.

Internal enterprise investigations are often more thorough in their discussion of possible causes for the occurrence, and could have more extensive coupling with technical and organizational circumstances.

Thus, the ideal situation for external learning is that both types of investigations are available; then using both investigations may provide a better understanding of all mechanisms.

For internal enterprise learning, it will often be the internal investigation report which gives the most direct input to enterprise internal improvement.

3.7 Investigation Modelling Methods

An accident model serves as a simplified representation of the real-world processes that result in the accident loss (Kjellén and Albrechtsen 2017). So far, various accident models have been developed and can be classified into three main types:

(1) Sequential models, which conceptualize an accident as a progression of events in a defined sequence.
(2) Epidemiological models, which draw parallels between the evolution of an accident and the transmission of a disease, suggest that the accident arises from a combination of both observable and hidden factors that coexist within a specific spatial and temporal context.
(3) Systemic accident models, which aim to illustrate the performance at the systemic level, view the accident as an emergent phenomenon resulting from the complex and nonlinear interaction among the system components.

Each of these models possesses its own characteristics and limitations based on its own interpretation of the accident phenomenon. This section introduces two widely used methodologies in practice, namely sequential timed events plotting (STEP) and man-technology-organization (MTO), both of which are classified as sequential models.

3.7.1 Sequential Timed Events Plotting Method

The STEP method, developed by Hendrick and Benner (1987), is grounded in the event sequence model and provides a structured process for accident investigation. By emphasizing the multi-linear sequences of events, the method adopts a process-oriented view of the accident phenomena. It conceptualizes an accident as a process that begins with an undesirable deviation within the system and concludes with a harmful event. The subsequent section provides a concise overview of the STEP method as detailed in the ROSS report (Sklet 2002).

Figure 3.9 illustrates a typical structure of a STEP diagram, which consists of the following main elements:

- An event is defined as a single action performed by one actor.
- An actor may refer to either an individual or an entity that has a direct influence on the flow or events that constitute the accident process.
- An action is an activity performed by the actor.
- An arrow illustrates the logical relationships between two or more events.
- The horizontal axis represents the timeline for documenting the beginning and ending of events. The timeline is developed for all actors involved in the accident.
- The vertical axis represents actors. Note that the role of the horizontal and vertical axes can be interchanged.

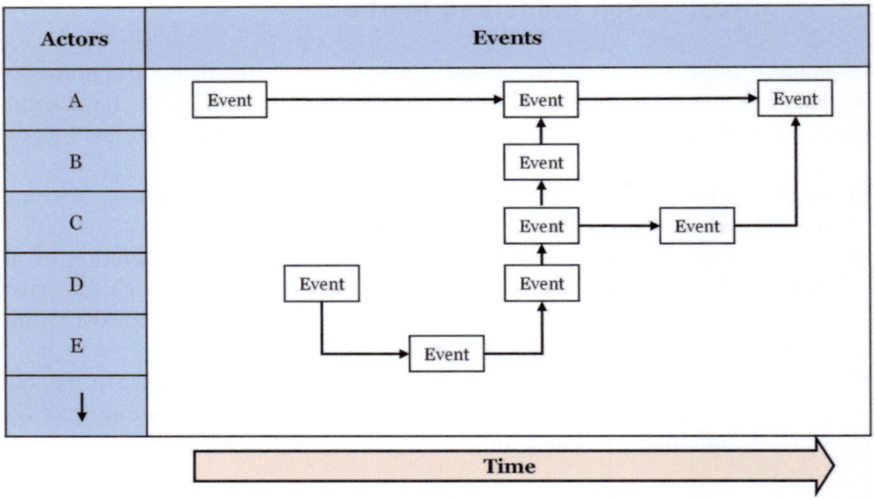

Fig. 3.9 The schematic representation of a STEP diagram

When analyzing each event presented in the STEP diagram, it is advisable to consider the question: "Do the preceding actions sufficiently trigger the subsequent actions (or events), or are additional actions required?" If a prior action is deemed essential for the occurrence of a subsequent event, an arrow ought to be added connecting the preceding event and the subsequent event to indicate this relationship.

In addition, it is essential to investigate the events that occurred during a gap where specific evidence is unavailable. The Back-STEP technique allows for a reverse analysis, beginning with the event on the right side of the gap and moving towards the left event. Throughout the Back-STEP process, a series of questions regarding "what could have contributed to it" is posed. Note that the answers may result in multiple pathways connecting the events on either side of the gap.

It is important to highlight that the STEP diagram places less emphasis on identifying the causes of accidents. Instead, the STEP method emphasizes the importance of identifying safety problems as well as developing safety recommendations based on findings from accident investigations.

3.7.2 Man-Technology-Organization Analysis

The MTO-analysis has been designed as a technique primarily for investigating accidents, utilizing the event causation and sequencing model. Its prototype is derived from the human performance enhancement system (HPES) developed in the nuclear sector. Currently, the MTO-analysis is extensively employed by Havtil for investigating accidents on the NCS. Its foundation is laid by the principle that human, technical and organizational factors hold equal importance in the investigation of accidents. The subsequent section provides a concise overview of the MTO-analysis as detailed in the SINTEF report (Tinmannsvik et al. 2005).

Figure 3.10 illustrates a typical structure of the MTO-analysis diagram, which consists of three main components:

(a) Conducting a structured analysis through the use of an event and causal factor (ECF) diagram. The sequence of events is developed horizontally and illustrated by a block diagram. Subsequently, potential human, technical, and organizational causes for each event are identified and arranged vertically for the corresponding event.
(b) Performing a change analysis by detailing the ways in which the events and conditions have deviated from the normal situation. Both the normal situation and the deviation are presented.
(c) Performing a barrier analysis to identify technical, operational, and organizational barriers that have either failed or are absent. The failed or absent barriers will be illustrated beneath the corresponding event.

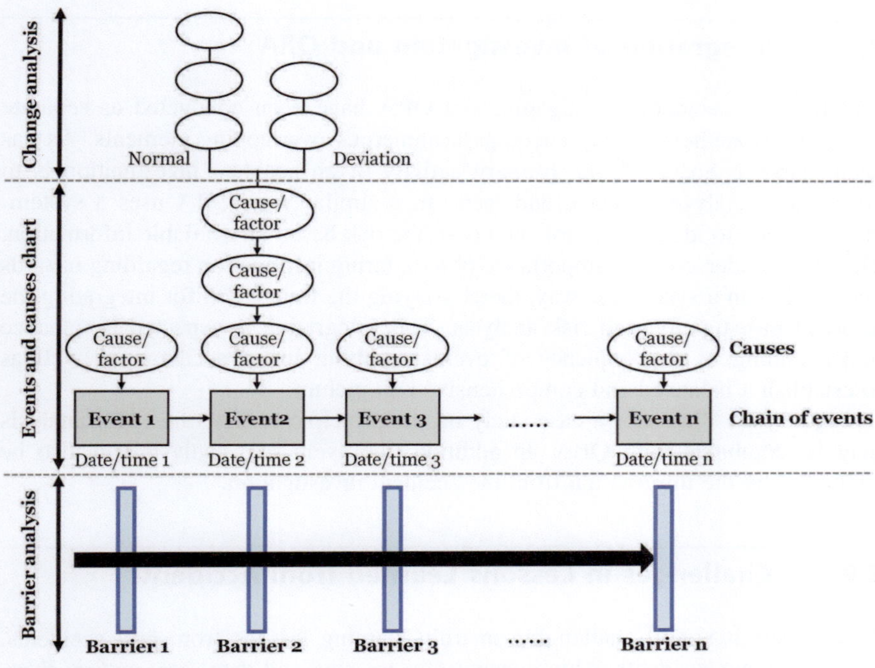

Fig. 3.10 The typical structure of an MTO diagram. *Source* Adapted from Vinnem and Røed (2020)

Further, it is essential to make practical and detailed recommendations within the MTO-analysis, which facilitates the identification of suggestions from technical, operational, and organizational viewpoints.

The MTO-analysis also includes a checklist for identifying the following types of failure causes. Note that for each failure cause, a further detailed checklist is prepared for outlining the basic causes.

(1) Technical failure,
(2) Working environment,
(3) Operational organization,
(4) Routines for change management,
(5) Installation management,
(6) Human-machine interface (HMI),
(7) Working schedule,
(8) Communication,
(9) Procedures/Instructions,
(10) Supervision,
(11) Working practice,
(12) Competence/Training.

3.8 Integration of Investigation and QRA

Traditionally, accident investigation and QRA have been conducted as separate activities. Nonetheless, they encompass numerous overlapping elements. As has been noted in Sect. 3.5, the primary activity of an accident investigation is to collect and analyze evidence and facts. In a similar way, QRA uses a systematic approach to identify hazards and describe risk based on available information. They both underscore the importance of structuring information regarding hazards and barriers in a systematic way, thereby laying the foundation for integrating the accident investigation and risk analysis. This integration is expected to enhance understanding of the sequence of events resulting in an accident, as well as to establish a balanced and comprehensive risk picture.

Figure 3.11 illustrates a case study on the principle of how the MTO-analysis may be combined with QRA. In addition, the event tree analysis can also be supported by the information from the accident investigation.

3.9 Challenges in Lessons Learned from Accidents

To this day, it is still challenging in truly learning lessons from past accidents. The analogous accidents, which occurred every now and then, can confirm these challenges. The following challenges have been clearly pointed out by Leveson (2011, 2020).

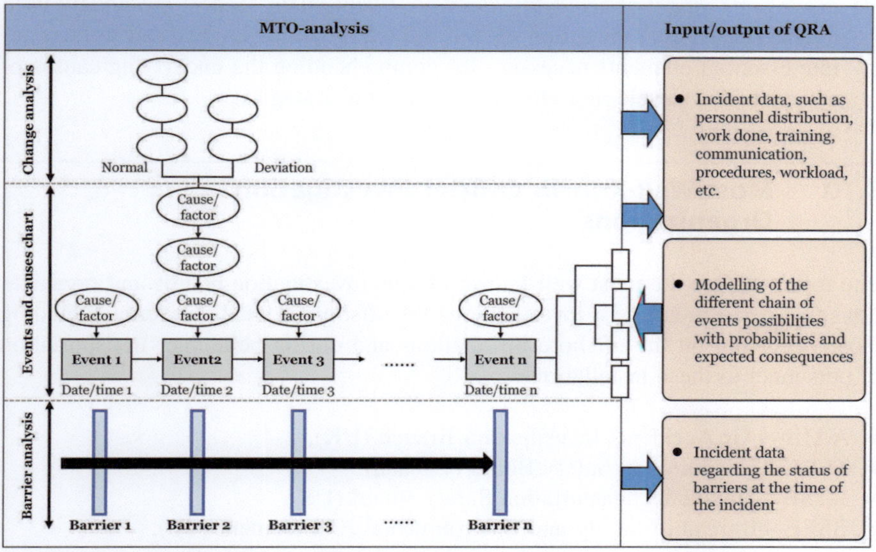

Fig. 3.11 The combination principle between the MTO-analysis and QRA, adapted from Skogdalen and Vinnem (2012) with Elsevier permission

(1) *Root cause seduction.* Organizations usually prioritize technical failures and operator errors, neglecting systematic and managerial factors. This oversight can lead to a complex scenario resembling a sophisticated "whack-a-mole" game, where symptoms are identified and addressed, yet the underlying processes that give rise to these symptoms remain unexamined.
(2) *Oversimplification of causes.* The attribution of causes is often limited to operator error, managerial deficiencies, inadequacies in equipment design, issues related to safety culture and regulatory shortcomings. However, the fundamental flaws within the safety control framework are frequently overlooked.
(3) *Hindsight bias.* After an accident, it becomes relatively straightforward to identify where individuals made mistakes, what actions they should have taken or avoided, and to recognize critical information that was overlooked. Such clarity is challenging to achieve prior to the occurrence of the event.
(4) *Confirmation bias.* Individuals tend to seek out evidence that supports their own goals, adhering to the notion that "what you look for is what you find".
(5) *Focus on blame.* When the focus shifts to assigning blame, the investigation often halts upon finding a suitable target for blame. Blame is the enemy of safety.
(6) *Narrow view of human error.* There is a prevailing assumption that the operator error is considered to be the main cause of most accidents, often overlooking the significant impact of the working environment on the operator's performance.

(7) *Inadequate model of accident causation.* Event-chain models are widely used today. These linear causation models are overly simplistic and fail to incorporate essential elements necessary for comprehending the underlying causes of accidents and developing effective prevention strategies.

3.10 Most Well-Known Official Investigation Organizations

The following are the most well-known official investigation boards and organizations that perform accident investigations for offshore-related accidents, including those occurring on the offshore installations and during helicopter transportation of personnel to these installations.

- AAIB—Air Accidents Investigation Branch (UK),
- TSB—Transportation Safety Board (Canada),
- NTSB—National Transportation Safety Board (US),
- BSEE—Bureau of Safety and Environmental Enforcement (US),
- Havtil—Norwegian Ocean Industry Authority (Norway),
- NSIA—Norwegian Safety Investigation Authority (Norway),
- HSE—Health and Safety Executive (UK),
- NOPSEMA—National Offshore Petroleum Safety and Environmental Management Authority (Australia),
- CNSOPB—Nova Scotia Offshore Petroleum Board (Canada),
- DMA—Danish Maritime Authority (Denmark),
- ANP—National Agency for Petroleum, Natural Gas and Biofuels (Brazil).

3.11 Questions

Q1: What are the pros and cons of the STEP method and the MTO method, respectively?
Q2: Can the two objectives of investigating accidents mentioned in Sect. 3.5 be in conflict with each other?
Q3: Please propose the structure (i.e., main chapters) of an investigation report.
Q4: How can you, in practice, implement the integration of QRA studies into an accident investigation report?
Q5: What are the most relevant organizational levels for learning from accidents?

References

ANP (2016) Investigation report of the explosion incident that occurred on 11/02/2015 in the FPSO Cidade de São Mateus

BP (2010) Deepwater horizon: accident investigation report

BP (2022) Gulf of Mexico commitment. https://www.bp.com/en_us/united-states/home/community/gulf-commitment.html

Cullen L (1990) The public inquiry into the Piper Alpha disaster. Department of Energy. HMSO, London

Graham B, Reilly WK, Beinecke F, Boesch DF, Garcia TD, Murray CA, Ulmer F (2011) Deep water: the gulf oil disaster and the future of offshore drilling. Report to the President. Washington DC, USA

Havtil (2013) Hydrocarbon leak on Oseberg A on 17 June 2013. Investigation report no: 001053028

Hendrick K, Benner LJ (1987) Investigating accidents with STEP. Marcel Dekker, New York

HSE (2001) Root causes analysis: literature review. WS Atkins Consultants Ltd.

Khan F, Abbasi SA (1999) Major accidents in process industries and an analysis of causes and consequences. J Loss Prev Process Ind 12:361–378

Kjellén U, Albrechtsen E (2017) Prevention of accidents and unwanted occurrences: theory, methods, and tools in safety management. Taylor & Francis Group, LLC

Leveson N (2011) Engineering a safer world. MIT Press, Cambridge, MA

Leveson N (2020) How to learn more from accidents. 2020 MIT STAMP workshop. July 20 to August 7, 2020

NOU (1981) Alexander L. Kielland accident. NOU report. Oslo, Norway

OLF (2012) Deepwater horizon—lessons learned and follow up. Summary report. Stavanger, Norway

Paradies M, Busch D (1988) Root cause analysis at Savannah River plant. In: IEEE conference on human factors and power plants

Sklet S (2002) Methods for accident investigation. ROSS report 200208. Norwegian University of Science and Technology, Trondheim, Norway

Skogdalen JE, Vinnem JE (2012) Combining precursor incidents investigations and QRA in oil and gas industry. Reliab Eng Syst Saf 101:48–58

SOA (2012) Investigation report with respect to Penglai 19-3 oil spill accident. Final report. China

Tinmannsvik RK, Sklet S, Jersin E (2005) Investigation methodology: man, technology, organisation (in Norwegian only). SINTEF report no: STF38 A04422

Tinmannsvik R et al (2011) The Deepwater horizon accident: causes, lessons learned and recommendations for the Norwegian petroleum activity, Executive summary. SINTEF report, May 2011

US DOE (1999) Conducting accident investigations. Tech. report. US Department of Energy, Washington, DC

Vinnem JE (2018) FPSO Cidade de São Mateus gas explosion– Lessons learned. Safety Science, 101:295–304

Vinnem JE, Røed S (2020) Offshore risk assessment: principles, modelling, and applications. Fourth Edition, Springer-Verlag London Ltd

Yue LQ, Chen XD, Yue PS, Shi Q, Zhang MY, Tang ZJ, Zhang CH (2018) Considerations on the warnings of Penglai 19-3 oil spill accidents to the Bohai resource development. In: Proceedings of the 2nd academic conference on the ocean development and management, Suzhou, China, pp 170–176

Fourth Lesson: Main Techniques for Risk Modelling

4

Use models by all means if you find them useful but do not become a slave to them.

—Trevor Kletz (825)

Abstract

Risk modelling techniques are crucial tools for conducting a QRA. This lesson dives into the commonly used techniques in engineering practice, which facilitate the identification and calculation of risks. By applying these techniques, readers can acquire proficiency to effectively connect theoretical concepts with practical applications in risk analysis. The lesson begins with hazard identification (HAZID) techniques, including HAZOP (Hazard and Operability Study), PHA (Preliminary Hazard Analysis) and FMECA (Failure Modes, Effects and Criticality Analysis). Subsequently, the focus shifts to causal, probability and frequency analysis, where techniques like CED (Cause-and-Effect Diagram), FTA (Fault Tree Analysis) and BNs (Bayesian Networks) are explored. For risks influenced by HOFs, the specialized methodologies of BORA (Barrier and Operational Risk Analysis) and Risk_OMT (Risk Modelling: Integration of Organizational, Human and Technical factors) are introduced. With respect to consequence analysis, ETA (Event Tree Analysis), ESD (Event Sequence Diagram) and escalation modelling are presented. Finally, the lesson discusses the principle of online risk analysis, which draws particular focus from regulatory authorities.

X. Zhen and J.-E. Vinnem, *Ten Lessons on Risks*,
https://doi.org/10.1007/978-3-031-99064-9_4

4.1 Hazard Identification

In the Second Lesson of this book, the importance of hazard identification, commonly known as HAZID, is underscored. This section outlines several widely acknowledged HAZID techniques, including HAZOP, PHA and FMECA, addressing their objectives, analysis procedure, worksheet, as well as strengths and limitations.

4.1.1 Hazard and Operability Study

HAZOP represents a structured and systematic technique for examining a defined system to identify hazards and potential operability problems. This technique is qualitative and is executed by a team of experts from various disciplines. HAZOP was originally developed for use in chemical process plants, and has been extended to a wide range of scenarios where interactions occur between hardware, software operators, and the external environment. Notably, HAZOP has also been extensively employed in the planning and development of offshore petroleum facilities. The following introduction is primarily based on the works of Mannan (2012), Crawley and Tyler (2015), and the international standard IEC 61882 (IEC 2016).

The fundamental principle of HAZOP involves the application of a set of guidewords in conjunction with system parameters to identify possible deviations from design intention, determine the probable causes and consequences of these deviations, and assess the barriers in place to prevent such deviations. The design intention serves as a conceptual representation outlining the anticipated developments and the essential parameters that will be investigated. Then, the system is typically divided into a manageable set of "study nodes" to which the guidewords can be applied and the design intention can be evaluated.

The foundation of HAZOP is laid by all the relevant and available materials, including the design documents, operational procedures, previous hazard study reports, etc. The validity of HAZOP rests with the HAZOP team's expertise and experience, the available information's precision, as well as the design quality. Although it's recommended to carry out a HAZOP study as early as possible during the design phase to allow for design modifications, it is crucial to recognize that it is not a substitute for a good design. If HAZOP uncovers an excessive number of design deficiencies, there could be something fundamentally wrong within the design framework.

HAZOP Objectives

HAZOP is conducted with the objectives of:

- identifying and evaluating any residual hazards associated with a planned process or operation that may not have been identified or eliminated during early phases;
- identifying critical operability problems related to the system and maintenance operations.

- recommending supplementary actions to mitigate the identified problems.

HAZOP Analysis Procedure

A comprehensive HAZOP analysis procedure mainly encompasses the following steps:

STEP 1: Planning and preparation.
A full description of the design and critical parameters should be developed to ensure the HAZOP team has a full understanding of the system. Then, a comprehensive design intention is formulated and recorded, enabling the HAZOP team to identify any situations that may fall outside the operational envelope. Further, the system is divided into a manageable set of study nodes.

STEP 2: Generating a deviation.
Each study node is examined, and meaningful deviations can be generated through the combination of guidewords and parameters. Table 4.1 lists some generic guidewords.

STEP 3: Identifying causes.
In instances where significant deviations occur, it is essential to identify their potential causes. Notably, if the consequences of the deviations are deemed negligible, there is no necessity to investigate further causes.

STEP 4: Evaluating consequences.
A thorough evaluation of the primary consequences of the identified deviations is crucial. This evaluation aids in understanding the progression of these deviations over time.

STEP 5: Identifying existing barriers.
Identification of existing barriers is necessary so as to be capable of proposing further measures for risk reduction.

STEP 6: Assessing risk.

Table 4.1 Commonly used HAZOP guidewords

Guideword	Means
NO/NOT	The design intention is not realized
MORE	There is a measurable increase in a specific parameter
LESS	There is a measurable decrease in a specific parameter
AS WELL AS	An additional activity takes place
PART OF	Only a part of design intention is realized
REVERSE	Logical opposite of design intention takes place
OTHER THAN	There is a total replacement
BEFORE/AFTER	The sequence of steps is disrupted
EARLY/LATE	The timing deviates from the design intention
FASTER/SLOWER	The step is done/not done with the right timing

Risk assessment is carried out for each deviation, which entails estimating both the frequency and the consequence of each deviation occurring. A risk priority number (RPN) is commonly calculated by Eqs. (4.1) and (4.2). Normally, the addition rule or the multiplication rule is employed when the frequency and consequence are measured with a logarithmic scale and linear scale, respectively.

$$\text{RPN} = \log(\text{consequence}) + \log(\text{frequency}) = S + L \tag{4.1}$$

$$\text{RPN} = S \times L \tag{4.2}$$

RPN is defined to evaluate the relative importance of the corresponding deviations; however, it lacks a clear interpretation. It is also a common practice to input the frequency of occurrence and severity of consequences with respect to each deviation into a risk matrix, thereby enabling the comparison of the deviation's risk against the established RTC. This contributes to evaluating the further measures for risk reduction. Notably, this step is not universally implemented and is considered optional.

STEP 7: Recommending actions.
If the existing barriers are found to be inadequate, possible actions will be recommended to either prevent the occurrence of the deviation or mitigate its potential consequences.

STEP 8: Report.
The HAZOP results should be fully recorded in a way that it is understandable to non-members of the HAZOP team, such as the managers. Typically, the HAZOP results are displayed in a designated worksheet.

HAZOP Worksheet

In order to record the analysis results in a systematic manner, a specialized HAZOP worksheet is developed, and Fig. 4.1 provides an example for reference.

It is important to note that there is no consensus on a standardized format for the HAZOP worksheet. Consequently, the HAZOP team has the flexibility to develop

Study Node:

Deviation:

Type: Drawings:

Design Conditions/Parameters: Equipment:

No.	Cause	Consequence	Barrier	Risk			Risk reducing measure	Responsibility	Comment
				Freq.	Cons.	RPN			
1									
2									
3									
4									
5									

Fig. 4.1 The format of a typical HAZOP worksheet

Table 4.2 Main strengths and limitations of the HAZOP technique

Strengths	Limitations
• Effective in addressing both technical failures and human errors • Capable of synthesizing collective judgments through the integration of diverse background knowledge • Capable of identifying existing barriers and recommending additional barriers • Systematic and comprehensive, facilitating the identification of hazardous deviations • Highly flexible, allowing the complex process systems to be divided into manageable nodes • Suitable for systems that necessitate the interaction of various disciplines and organizations	• Heavily relying on the facilitation provided by the team leader, as well as the expertise and experience of the experts involved • Relying on guide words, which may not adequately be linked to potential failures • Time consuming, allowing for endless discussions of details • Probably dominated by a few experts • Focusing on process-related hazards, which may not be well suited for addressing other types of hazards • Restricted to single failure conditions • Focusing on qualitative analysis for all issues

their own worksheet tailored to the specific objectives, which may involve the addition or removal of certain columns.

Strengths and Limitations

Table 4.2 provides a summary of the primary strengths and limitations associated with the HAZOP technique.

4.1.2 Preliminary Hazard Analysis

PHA is a qualitative technique employed to identify hazards and potential hazardous events. Additionally, it serves to evaluate the risk associated with each identified hazardous event by determining its corresponding frequency and severity. The term "preliminary" from PHA is distinguished from "detailed", indicating that a detailed system design is not required to be available while hazard identification can be conducted during the earlier stages, such as the conceptual design and the front-end engineering design stages. This PHA feature contributes to determining the optimal design of facilities and equipment.

Therefore, PHA is typically employed for initial risk analyses in the early stages of a project. It also serves as a preliminary step for a detailed risk analysis or a complete risk analysis for an uncomplicated system.

PHA Objectives

PHA is conducted with the objectives of:

- identifying potential hazards and hazardous events that could result in accidents;
- identifying the main causes of each hazardous event;

- assessing and ranking the risk level of each hazardous event;
- recommending relevant measures for risk reduction.

PHA Analysis Procedure

A comprehensive PHA analysis procedure mainly encompasses the following steps:

STEP 1: Planning and preparation.
This initial step involves defining the system to be analyzed, encompassing a comprehensive description of the system, its boundaries, operational modes, environmental conditions, existing barriers, etc. It is a common practice to decompose the system into sub-systems to facilitate a thorough identification of potential hazards. Furthermore, all the necessary data are gathered, including the design-related data, operation-related data, process-related data, hazard-related data, etc.

STEP 2: Hazard identification.
All hazards in relation to the concerned subsystems and operations are identified. Furthermore, possible hazardous events arising from these identified hazards are defined.

STEP 3: Assessment of risk.
The risk of each hazardous event is assessed through its occurrence frequency and corresponding consequence. The underlying causes of each hazardous event and existing barriers are identified as well. RPN for each hazardous event may be calculated. It is also a common practice to input the frequency of occurrence and severity of consequences with respect to each hazardous event into a risk matrix.

STEP 4: Recommendations for risk reduction.
If existing barriers are deemed inadequate, potential measures for risk reduction are recommended.

STEP 5: Report.
The analysis process and results are fully recorded in a specific PHA worksheet, and are reported to the managers and other stakeholders.

PHA Worksheet

Figure 4.2 presents an example of a PHA worksheet for reference. There exists no single, standardized format for the PHA worksheet. The PHA team can develop their own worksheet based on their objectives by adding or deleting some of the columns.

Strengths and Limitations

Table 4.3 provides a summary of the primary strengths and limitations associated with the PHA technique.

System: Analysts:

Subsystem/Operation: Date:

No.	Hazard	Hazardous event	Cause	Consequence	Risk			Risk reducing measures	Responsibility	Comment
					Freq.	Cons.	RPN			
1										
2										
3										
4										
5										

Fig. 4.2 The format of a typical PHA worksheet

Table 4.3 Main strengths and limitations of the PHA technique

Strengths	Limitations
• Relatively simple to implement • Applicable during the early design phase, enabling cost-effective modifications • Capable of providing a list of hazards as well as the occurrence frequency and the corresponding consequence of the hazardous event • Capable of providing useful input for the majority of risk analyses	• Difficult to illustrate the interaction effects between different hazards • Difficult to illustrate the effects of barriers

4.1.3 Failure Modes, Effects and Criticality Analysis

FMECA builds upon the foundation of Failure Modes and Effects Analysis (FMEA). It represents a systematic technique employed to examine a system, aimed at identifying potential failure modes, their underlying causes, as well as the consequent effects on the system's performance.

In the FMECA, the term "failure mode" can be defined as the way an item fails. The symbol "C" in the FMECA indicates that the criticality of the various failure effects is taken into account and ranked. FMECA is commonly conducted in the conceptualization and preliminary design phases of a system, enabling the identification and mitigation of potential failures in accordance with design changes. The following introduction is primarily informed by the international standard IEC 60812 (International Electrotechnical Commission 2006a).

FMECA Objectives

FMECA is conducted with several key objectives of:

- identifying the system's failure modes and their unwanted effects;
- identifying the underlying causes of the failure modes;
- determining the detection method for each failure mode;
- estimating the occurrence frequency and the consequence in association with each failure mode;
- evaluating the risk in relation to each failure mode;

- recommending measures for risk reduction.

FMECA Analysis Procedure

A comprehensive FMECA analysis procedure mainly encompasses the following steps:

STEP 1: Planning and preparation.
The initial step involves defining the system that will be analyzed, encompassing main system functions, system boundaries, operating conditions, environmental conditions, and other relevant aspects. All the relevant data are collected.

STEP 2: System structure analysis.
In this step, the system is decomposed into subsystems. The items at the bottom of the system hierarchy are typically designated as components, as depicted in Fig. 4.3. It is worth noting that the definition of failure modes, causes, and effects is contingent upon the specific level of the system structure. To be specific, failure effects identified at lower levels may manifest as failure modes at higher levels, while lower-level failure modes may serve as causes for failures at higher levels.
To facilitate this analysis, a functional block diagram (FBD) is constructed, which delineates the system structure, as depicted in Fig. 4.4. Each functional block comprises three basic elements: input, internal states, and outputs. The internal states are defined as either the system's ability to execute its designated function or its inability to achieve that. The output is determined based on the inputs and the specific state of the functional block, with interconnections represented by directional arrows.

STEP 3: Identification of failure modes, causes and effects.

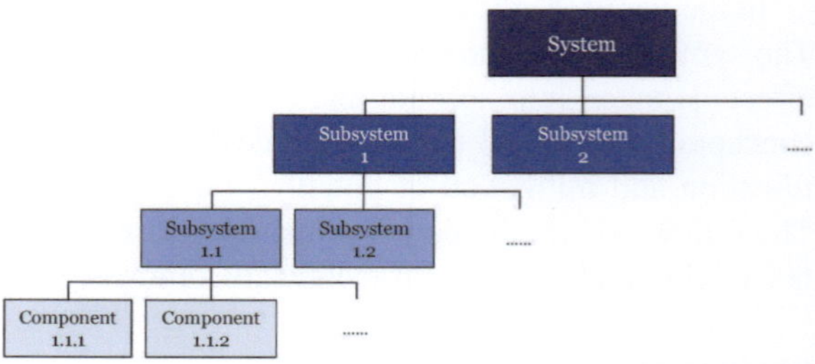

Fig. 4.3 System breakdown structure

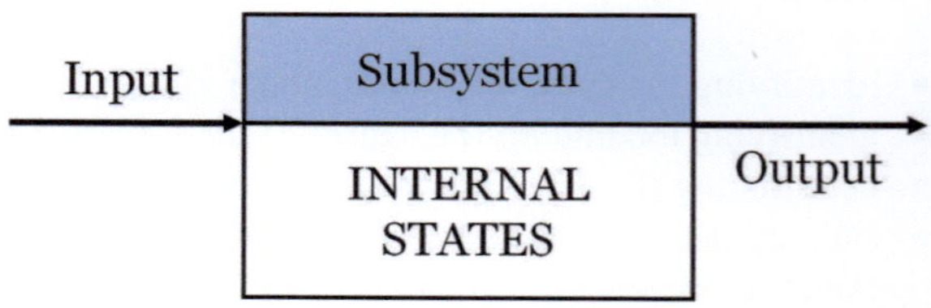

Fig. 4.4 The basic principle of the functional block

The functions and operational modes of each item in the system should be understood and documented by the analysts. Subsequently, the corresponding failure modes are identified. What counts is evaluating all items at the bottom within the established system boundaries.

The underlying causes associated with each failure mode are systematically identified and articulated. In particular, a single failure mode may be attributed to multiple causes; therefore, it is essential to determine the most likely independent causes. Furthermore, the identification and description of failure causes should be informed by the corresponding failure effects. In instances where failure modes exert negligible or minimal impact on the system functionality, it is unnecessary to ascertain the causes.

It is essential to identify and assess the consequences of failure modes, specifically the failure effects on the system performance. A failure effect may arise from one or multiple failure modes across various system components. The failure effects can be categorized into local effects and final effects. The local effect refers to how the failure mode affects the specific system item being analyzed, while the final effect describes the failure's impact on the overall system at its highest level. The identification of local effects serves as a foundation for sound decision-making when evaluating current alternative provisions or suggesting corrective measures.

STEP 4: Determination of the failure detection methods.

For each failure mode, it is essential to determine the method for detecting such failures. Failure detection may be implemented by a built-in-test or a special checkout procedure. The detection process should take into account the likelihood, timing, and location of potential failures.

STEP 5: Assessment of the risk.

The occurrence frequency of each failure mode and the severity of its consequence are estimated. Risk is usually presented by a risk matrix or an RPN, as defined in Eq. (4.3). The risk ranking is utilized to prioritize possible failure modes for mitigation actions and resource allocation.

$$\mathrm{RPN} = S \times O \times D \tag{4.3}$$

where

S represents the severity ranking of the failure mode; O represents the occurrence frequency ranking of the failure mode; D signifies the likelihood ranking of detecting and mitigating the failure prior to its impact on the system.

STEP 6: Recommendations of measures for risk reduction.

Potential measures for risk reduction are recommended if risk evaluation results indicate a need for actions.

STEP 7: Report.

The process and results of FMECA analysis are comprehensively recorded in a specific FMECA worksheet.

Item: Analysts:
Operating period: Date:

Item ref.	Item description and function	Failure mode	Failure causes	Local effect	Final effect	Detection method	Risk				Risk reducing measures	Responsibility	Comment
							O	S	D	RPN			

Fig. 4.5 Example of the format of an FMECA worksheet

Table 4.4 Main strengths and limitations of the FMECA method

Strengths	Limitations
• Highly structured and reliable • Easy to learn and understand • Suitable for complex systems	• Easy to be time-consuming • Unsuitable for analyzing multiple simultaneous failures • Highly dependent on the skills and experience of the analysts

FMECA Worksheet

Figure 4.5 presents an example of a format for an FMECA worksheet for reference. The FMECA team can also develop their own worksheet based on their objectives by adding or deleting some of the columns.

Strengths and Limitations

Table 4.4 presents a comprehensive overview of the primary strengths and limitations associated with the FMECA technique.

4.2 Causal, Probability and Frequency Analysis

The causal, probability and frequency analysis techniques are employed in QRA, with the objective of determining the following parameters:

- the causes of hazardous events;
- the relationship between the hazardous event and root causes;
- the frequency at which hazardous events occur;
- the importance of each cause in relation to the frequency of hazardous events;
- the probability of adverse operational and environmental conditions;
- the probability of a specific severe accident consequence.

There are many quantitative techniques for calculating the probabilities and frequencies of the hazardous event occurring. This section presents an introduction to the commonly used techniques, encompassing the CED, FTA and BNs. Additionally, a brief introduction to Dynamic Fault Tree Analysis and Dynamic Bayesian Networks is also encompassed.

4.2.1 Cause-and-Effect Diagram

CED is also known as the fishbone diagram due to its visual appearance or the Ishikawa diagram in recognition of its inventor (Ishikawa 1976). It serves as a purely qualitative and graphic tool for identification, categorization, and description of potential causes associated with a specified hazardous event. This tool is based on collaborative brainstorming among a team of analysts, and thus, its theoretical basis is rather limited.

Figure 4.6 illustrates the process for constructing a typical CED, and the following main steps are normally carried out.

STEP 1: The hazardous event is identified and defined within a box positioned on the right-hand side of CED. A horizontal line is drawn, pointing to the box. The box and the horizontal line symbolize "the head and spine of the fish", respectively.

STEP 2: The primary categories of causes contributing to the hazardous event are determined and drawn as "fishbones" positioned above and below the "spine of the fish". These categories of causes may encompass equipment, environment, methods, materials, and human factors.

STEP 3: For each identified category of causes, the contributing factors are identified through brainstorming sessions and drawn as "branch bones" extending from the "fishbone".

STEP 4: For each contributing factor, the sub-factors and further levels of contributing factors could be identified and drawn until the requisite level of detail is reached. The relationship between the specified hazardous event and all associated factors is thus established.

Upon constructing a CED, the logic and completeness of the diagram ought to be checked first. Subsequently, a qualitative analysis of the diagram is conducted, aimed at identifying and prioritizing the most probable causes of the hazardous event. In particular, it is important to find out any identical or similar causes that may appear in multiple locations within the diagram.

Table 4.5 provides a summary of the primary strengths and limitations associated with the CED.

4.2.2 Fault Tree Analysis

FTA is a logical and structured approach employed for identifying underlying causes behind the system failures, including the causes of hazardous events as well as barrier failures. It can be employed for both qualitative and quantitative causal analysis in risk studies. FTA is represented through a graphical model that demonstrates all potential failure combinations, which could lead to the occurrence of the hazardous event. The International Standard (International Electrotechnical Commission, 2006b) provides guidance on the description and application of FTA.

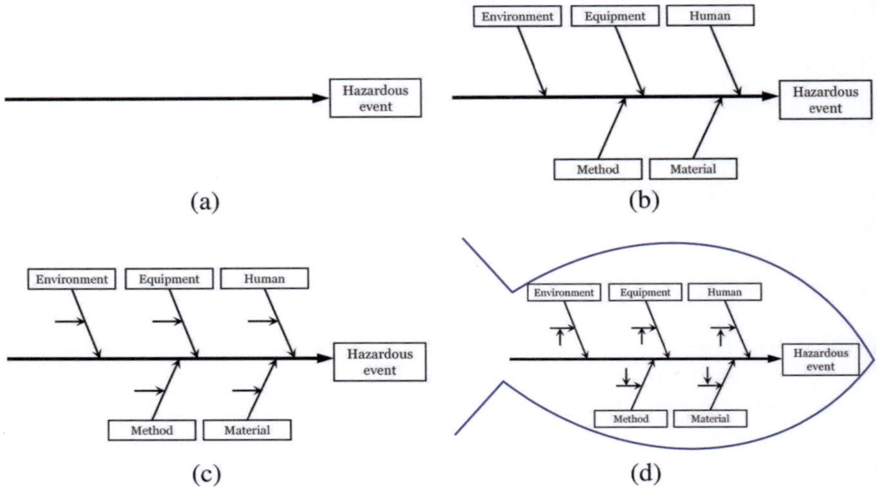

Fig. 4.6 The construction process of a CED

Table 4.5 Main strengths and limitations of the CED

Strengths	Limitations
• Suitable for collaborative brainstorming • Easy to learn and understand • Capable of organizing and determining causes • Increasing the knowledge of process	• Purely qualitative in nature • Not ranking causes • Can become complicated • Lack of theoretical foundation

The significant advantage of FTA is its capacity to integrate both technical and non-technical failures, thereby determining and elucidating all the possible pathways to the hazardous event occurring.

FTA is conducted with the objectives of:

- determining the causes or combination of causes of the hazardous event;
- quantifying the probability of the hazardous event occurring;
- identifying measures to reduce the probability of the hazardous event occurring.

In contrast to the aforementioned approaches focused solely on the single failure, such as HAZOP, PHA and FMECA, FTA is suitable to analyze the complex and redundant systems with a focus on the combination of failures. Consequently, the starting point of an FTA is often the accomplished HAZOP, PHA or FMECA, which has determined the hazardous event that necessitates further investigation.

FTA is a deductive, top-down technique that involves the following key steps:

STEP 1: Definition of the system and scope of the analysis.

The system to be analyzed should be clearly defined through a description of its function as well as an identification of the system interfaces. This definition is crucial for achieving a thorough understanding of the system's working principle. The scope of the analysis should delineate the system's physical boundaries, its operational states, the external stresses to be considered, as well as the degree of detail in the analysis.

STEP 2: Construction of the fault tree.

During this step, it is essential to clearly define the TOP event, which signifies the hazardous event, by addressing its nature (what), location (where), and timing (when).

After that, it is essential to identify the immediate causal events and conditions leading to the TOP event.

Then, the immediate causal events are linked to the TOP event using logic gates, specifically AND-gates or OR-gates.

The above procedure proceeds until an appropriate level of detail is achieved, where the events are independent and classified as basic events. A summary of the commonly used FTA symbols is provided in Table 4.6.

STEP 3: Analysis of the fault tree logic.

The minimal cut sets (MCSs) are identified first. A cut set is defined as a collection of events that, if all occur, would lead to the occurrence of the TOP event. Consequently, the MCS signifies the smallest combination of events required to trigger the TOP event, indicating that if any event within this set fails to occur, the TOP event will not occur.

Then, the MCSs ought to be ranked based on their order, specifically according to the number of basic events they comprise.

Finally, the fault tree is analyzed quantitatively. The probabilities associated with the basic events need to be known in advance for calculating the probability of the TOP event in accordance with the following equations, provided that all minimal cut sets are independent.

$$Q_0 = \Pr(\text{TOP}) = \Pr(C_1 \cup C_2 \cup \cdots \cup C_k) \tag{4.4}$$

$$Q_0 = 1 - \prod_{j=1}^{k} (1 - \tilde{Q}_j) \tag{4.5}$$

$$\tilde{Q}_j = \Pr\left(E_{j,1} \cap E_{j,2} \cap \cdots \cap E_{j,n}\right) = \prod_{i \in C_j} q_i \tag{4.6}$$

where

Q_0 denotes the occurrence probability of the TOP event; C_k denotes the MCS; $\tilde{Q}_j$ denotes the probability that the C_j fails; q_i denotes the probability that the basic event $E_{j,i}$ fails within the C_j; n is the order of the C_j.

Table 4.6 Commonly-used FTA symbols

Item	Symbol	Description	Number of inputs
Logic gates	 OR gate	The output event is activated when any input event occurs	≥ 2
	 AND gate	The output event is triggered by the simultaneous occurrence of all input events	≥ 2
Input events	 Basic event	The lowest level event for which the probability of occurrence is available	0
	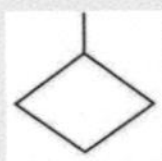 Undeveloped event	A primary event that signifies a component of the system that has not yet been fully developed	0
State description	 Comment rectangle	It serves as supplementary information	0
Transfer gates	 Transfer IN	The gate signifies that this particular component of the system has been developed in a different section or page of the diagram The designation "IN" indicates that the developed gate is located elsewhere, whereas "OUT" signifies that the gate developed in this location will be utilized in another context	0
	 Transfer OUT		0

STEP 4: Reporting on results of the analysis.

The FTA report primarily includes the following basic items, covering the objectives and scope, the system description, assumptions, the composition of the team, references for basic events and undeveloped events, as well as the analysis, results, conclusions, and recommendations.

Table 4.7 Main strengths and limitations of the FTA

Strengths	Limitations
• User-friendly and widely recognized • Providing a clear and coherent comprehension of system characteristics • Capable of identifying all causes of a TOP event • Capable of prioritizing risk-reducing measure	• Constrained to known failures • Time-consuming to construct • The fault tree can become excessively complicated • Challenging to update in response to changes

The conventional FTA in its static form is limited in its capacity to model the dynamic interactions among system components. To enhance the modelling capabilities of the conventional FTA, a dynamic FTA has been developed. This improvement primarily focuses on the logic gates utilized within the analysis. In the dynamic FTA, both Boolean logic gates and dynamic gates are employed to delineate the logical relationships between intermediate and basic events.

Therefore, the dynamic FTA is challenging to quantitatively determine the probability of the TOP event by use of traditional algebraic and cut-set based methods. As a result, alternative methodologies, such as the Markov method (Ross 1996), Petri Nets (Petri 1962), and BNs, among others, need to be harmoniously combined when determining the probability of the TOP event. This section will subsequently introduce the BNs. For a more in-depth exploration of dynamic FTA, the work conducted by Aslansefat et al. (2020) is recommended.

Table 4.7 provides a summary of the primary strengths and limitations associated with the FTA technique.

4.2.3 Bayesian Networks

BN is modeled as a directed acyclic graph (DAG), which represents the causal relationships between critical variables and their resultant outcomes. The DAG comprises a collection of nodes and directed arcs connecting these nodes. Each node signifies a particular state or condition, while a directed arc denotes a direct influence. Particularly, the term "acyclic" indicates that the presence of cycles within the BN is prohibited. For instance, Fig. 4.7 displays a simple BN, consisting of two nodes labeled "RIF (risk influencing factor)" and "Score", connected by a directed arc. In this context, the node designated as "RIF" is identified as a parent node to the node labeled "Score," which is consequently referred to as a child node of "RIF". The node "RIF," which lacks any parent nodes, is classified as a root node, whereas the node "Score," which lacks any child nodes, is termed a leaf node or end node. The BN analysis can be classified into two categories: qualitative and quantitative, based on the specific objectives of the investigation. For a more comprehensive introduction to BNs, the works of Jensen (2001), Pearl (2001) and Kjærulff and Madsen (2014) are recommended.

Fig. 4.7 Illustration of a simple BN

The BN analysis is conducted with the objectives of:

- determining the posterior probability distribution of outcome variables in light of new evidence;
- determining the causes and influencing factors of the outcome;
- ranking the priority of each influencing factor to the probability of the outcome.

In contrast to FTA, the Bayesian network approach facilitates measurements in multiple dimensions as well as enables the modelling of cause-and-effect relationships among components of a system that are not deterministic. In addition, the fault tree can be easily transformed into a Bayesian network.

The primary procedures associated with the BN analysis are outlined in the following:

STEP 1: Construction of the BN.

The initial step involves defining the leaf node(s). The hazardous event is typically defined as the leaf node within the BN.

Subsequently, a finite set of parent nodes is defined to represent the events and factors affecting the leaf node(s). The node is typically expressed through a random variable that follows a discrete distribution. Although a continuous distribution could be employed, it would complicate the BN analysis. The parent node states are denoted by the values of the random variables, with each variable potentially exhibiting two or more possible states.

Finally, the nodes are connected through a set of directed arcs. An arc originating from one node and terminating at another signifies a statistical dependency between the associated variables. Consequently, the arrow's direction signifies that the value taken by the child node is contingent upon the value taken by the parent node, thereby illustrating the direct influence exerted by the parent node on the child node.

STEP 2: Quantitative analysis of the Bayesian network.

This step focuses on establishing the conditional probability tables (CPTs) for the connected nodes. Table 4.8 presents the CPT of node B as depicted in Fig. 4.8. Taking into account the conditional dependencies among the variables, a BN characterizes the joint probability distribution $P(Z)$ of the variables $Z = \{Z_1, \ldots, Z_n\}$ as

$$P(Z) = \prod_{i=1}^{n} P(Z_i|Pa(Z_i)) \tag{4.7}$$

where
$P(Z_i|Pa(Z_i))$ denotes the conditional probabilities associated with different variables; $Pa(Z_i)$ denotes the set of parent nodes corresponding to each Z_i, where i ranges from 1 to n.
Then, the probabilities associated with the final outcomes can be computed. Additionally, the priority of each parent node could be determined through the sensitivity analysis, which identifies the variables exerting the most influence on the leaf node.

Notably, the technique of dynamic Bayesian networks (DBNs) has attracted attention due to its capacity to model the temporal dependencies and evolution of systems over time. DBNs represent an extension of the traditional BNs, and their core steps are outlined in the following.

STEP 1: DBN construction.
Initially, the BNs are constructed as previously stated. Subsequently, the dynamic nodes, such as the nodes D and E depicted in Fig. 4.9, are defined. These dynamic nodes can be characterized by dynamic variables, whose probability distributions evolve over time.
Then, the dynamic nodes in different time slices (Δt, $2\Delta t$, …, $n\Delta t$) are connected through a set of directed arcs. Time slices denote the discrete timestamps associated with the BNs. Within the framework for modelling a DBN, it is assumed that a node is influenced not solely by its parent nodes within the current time slice, but also by its parent nodes and itself from the preceding time slice, as illustrated in Fig. 4.9.

STEP 2: DBN analysis

Table 4.8 CPT of the node B

A	Happen		Not happen	
C	Happen	Not happen	Happen	Not happen
B (Happen)	$P(B|AC)$	$P(B|A\overline{C})$	$P(B|\overline{A}C)$	$P(B|\overline{AC})$
B (Not happen)	$1 - P(B|AC)$	$1 - P(B|A\overline{C})$	$1 - P(B|\overline{A}C)$	$1 - P(B|\overline{AC})$

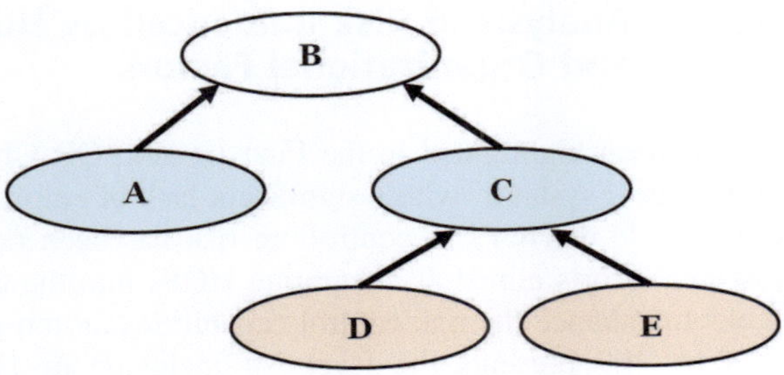

Fig. 4.8 A graphical representation of the Bayesian network

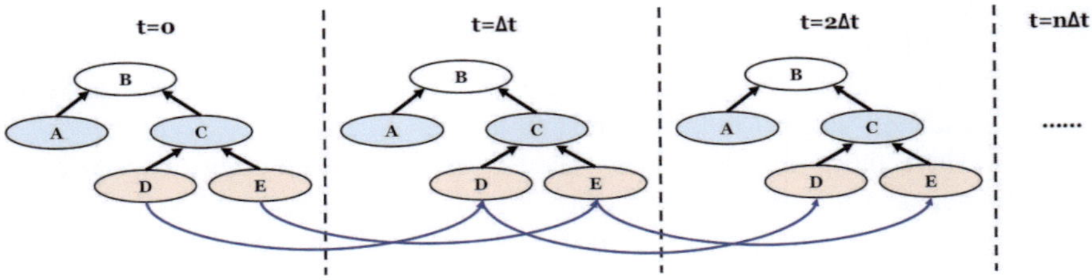

Fig. 4.9 Construction of a dynamic Bayesian network

In this step, the state transition probability tables for the dynamic nodes need to be defined. The state transition probability represents temporal conditional dependencies between two neighboring time slices.
The joint probability distribution of nodes within T time-slices can be determined by Eqs. (4.8) and (4.9). The DBNs can be interpreted by "unrolling" the temporal BNs consisting of two slices until the desired number of T time-slices is obtained.

$$P(Z_t|Z_{t-1}) = \prod_{i=1}^{n} P\left(Z_{i,t}|Pa\left(Z_{i,t}\right)\right) \tag{4.8}$$

$$P(Z_{1:T}) = \prod_{t=1}^{T}\prod_{i=1}^{n} P\left(Z_{i,t}|Pa\left(Z_{i,t}\right)\right) \tag{4.9}$$

where $Z_{i,\,t}$ denotes the ith node at time t; $Pa(Z_{i,\,t})$ denotes the parent node of $Z_{i,\,t}$ in the previous and the current time slices; $P(Z_t|Z_{t\,-\,1})$ denotes a variable transition model from the previous time slice to the current one within a DBN.

It is worth noting that several software packages, such as GeNie (BayesFusion 2025), Netica (Norsys 2025) and HUGIN (Hugin Expert 2025), have been developed for the construction and analysis of Bayesian networks.

4.3 Analysis of Risk Influenced by Human and Organizational Factors

As has been highlighted in the First Lesson, QRA has traditionally concentrated on technical systems, with a significant lack of emphasis on HOFs. Nowadays, the critical role of HOFs in controlling risk has been realized, leading to significant research efforts aimed at integrating HOFs into the QRA model. This integration seeks to enhance the risk control capabilities in non-technical factors.

Table 4.9 presents the levels of inclusion of HOFs in QRA. This section delineates two relevant methods at level 3, as proposed by the authors.

Table 4.9 Levels of HOFs inclusion (Skogdalen and Vinnem 2011)

Level	Characteristics
4	• The QRA is an integral component of the risk management system • The QRA results serve as the foundation for daily risk management practices • The QRA is recognized and endorsed at all organizational levels • The QRA presents the status of safety barriers in conjunction with risk indicators
3	• HOFs data is collected in a systematic way • QRA models are refined based on insights derived from HOFs • The identification of error causes is conducted to facilitate the development of risk-reducing measures • A systematic procedure is employed to generate reproducible qualitative and quantitative results
2	• The importance of HOFs is explained • The impact of HOFs on various components of the system is described to a certain extent • Human errors are determined in a separate way • Interviews with crew members are included. The results are commented on while the models and calculations remain unmodified
1	• An analysis of both technical and operational factors is performed • Risk-reducing measures encompass both technical and operational measures

- Barrier and operational risk analysis (BORA).
- Risk Modelling: Integration of Organizational, Human and Technical factors (Risk_OMT).

4.3.1 Barrier and Operational Risk Analysis

The BORA method (Vinnem 2007) was developed during the period between 2004 and 2006, with the primary aim of analyzing the risk of hydrocarbon leaks on offshore petroleum installations. Its purpose is to provide a modelling technique for operational hazards that allows for quantification and the possibility to visualize the impacts of operational improvements.

Figure 4.10 presents the fundamental components of a generic risk model for the BORA method. Building upon this framework, the principal modules and the associated steps of the BORA method are succinctly introduced in the remainder of this section.

Module 1 Qualitative Analysis of Scenarios, Barriers and RIFs.

- Identification of initiating events. This step marks the commencement of the analysis. Initiating events refer to technical failures or operational errors within the hydrocarbon system, which have the potential to result in a leak.

- Development of barrier block diagrams (BBDs). Typically, one or more barrier systems are established to prevent the progression of initiating events into undesirable events. These in-place barrier systems are modelled by BBDs, as illustrated in Fig. 4.11. If the barrier system operates as designed, any deviations will be detected and rectified, thereby preventing a leak from occurring. In practical applications, these BBDs can be converted into event trees.
- Modelling the performance of barrier systems. The performance of barrier systems is modeled by the use of the fault tree. Aiming at generalizing the fault tree, the TOP event is typically defined as "failure or degradation of the barrier system". The causes of the TOP

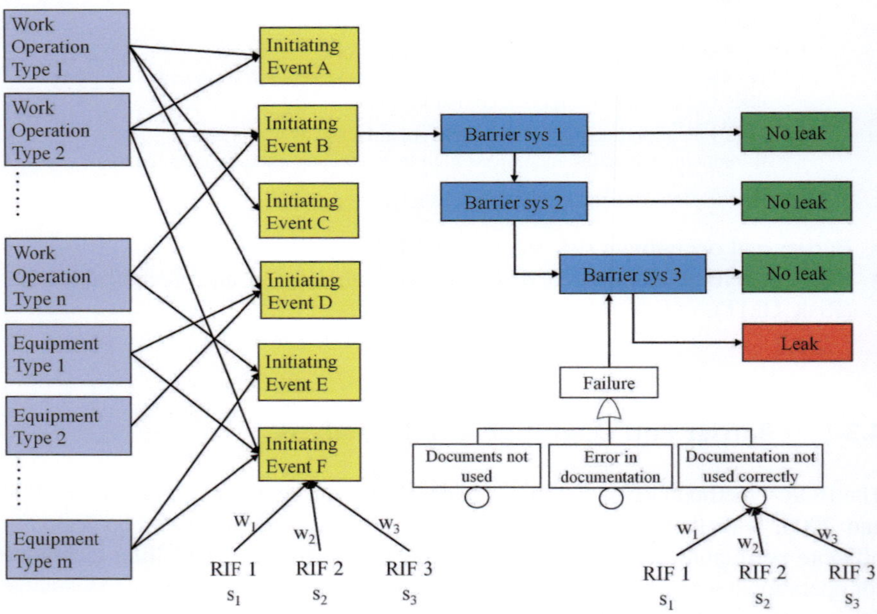

Fig. 4.10 Representation of a generic risk model for the BORA method

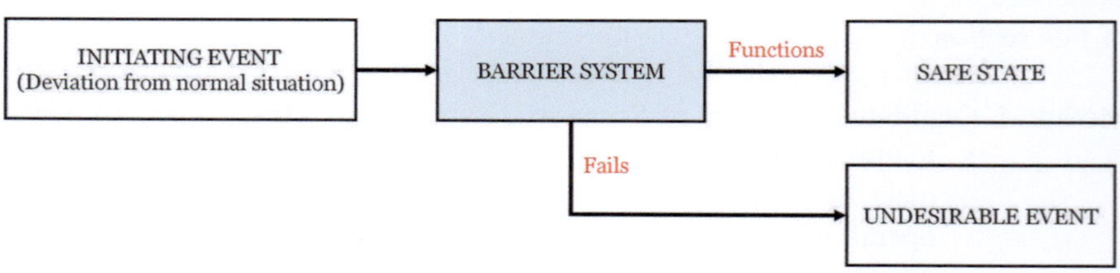

Fig. 4.11 The construction principle of a barrier block diagram

event are categorized into three primary groups: inadequate functionality of the barrier system, technical failures, and human errors, as illustrated in Fig. 4.12.

- Development of risk influencing diagrams (RIDs). RIDs are developed to identify and illustrate the RIFs that affect the occurrence probabilities or frequencies of the basic events represented in the fault trees. Identification of RIFs for each basic event is grounded in a generic framework, as illustrated in Fig. 4.13. A concise description of each RIF is provided in Table 4.10.

Module 2 Quantification of Industry Average Frequencies/Probabilities.

- Assignment of generic frequencies to the initiating events. Generic frequencies for initiating events can be derived from historical data, which reflects the distribution of the occurrence frequencies of the initiating events.
- Assignment of generic frequencies/probabilities to the basic events. To quantify the probability of barrier systems, generic input data is assigned to the pertinent basic events within the fault tree.
- Calculation of leak frequency utilizing the generic input data. The generic frequencies of leaks can be computed based on the assigned frequency of the initiating event, the failure probability of the barrier system, and the structure of BBD.

Module 3 Quantification of Specific Frequencies/Probabilities.

- Assignment of priority weights and scores to RIFs. The RIF is characterized by two key elements: "weight" and "score". The weight reflects the extent to which the RIF affects the probability of an event, while the score represents the current status of the RIF. To determine the weights of the RIFs, a five-point scale is utilized, assigning relative weights of 10-8-6-4-2. Subsequently, these weights are normalized. Then, each RIF is assigned a score ranging from A to F, with score A denoting the highest standard in the industry, score C representing the industry average, and score F indicating

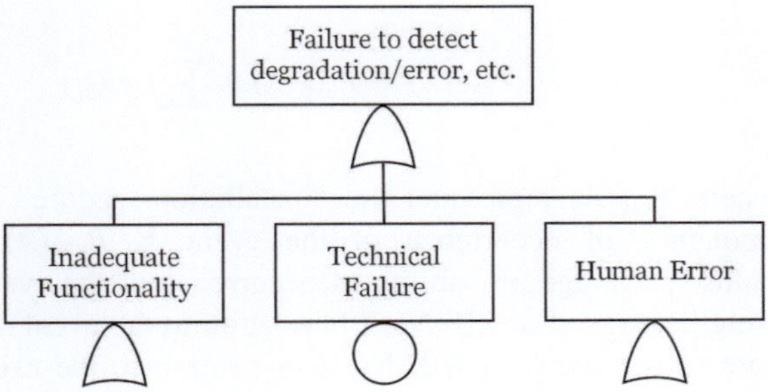

Fig. 4.12 Generic fault tree for analyzing the failure of barrier systems

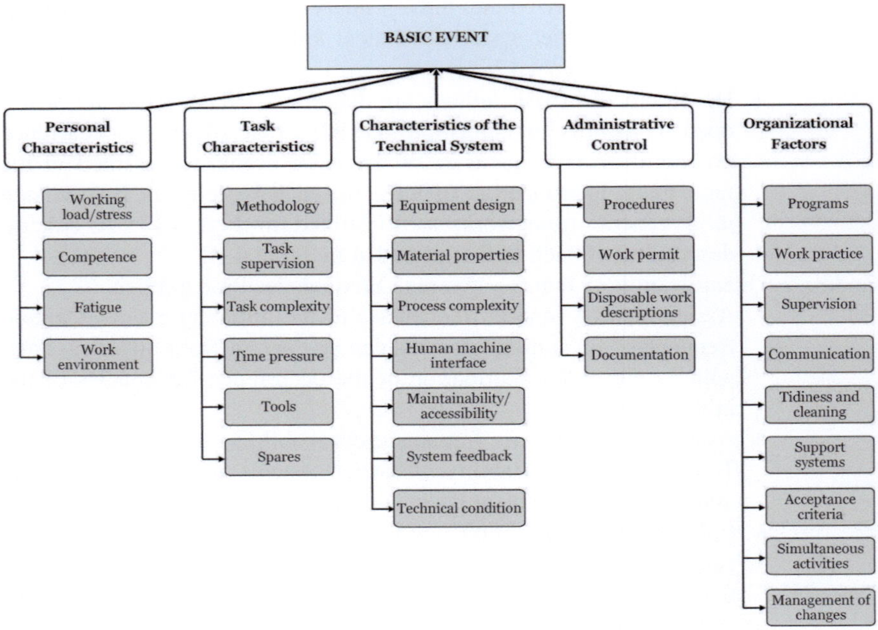

Fig. 4.13 Generic framework for identifying RIFs

the lowest standard of practice within the industry. Table 4.11 illustrates the general framework for scoring RIFs, employing a six-point scale.

- Adjustment of industry-average frequencies/probabilities. The average probabilities/frequencies in the industry are revised on the basis of the RIDs in accordance with a comprehensive assessment of both the weights and status of the RIFs. The adjustment principle is implemented, as presented in Eq. (4.10).

$$P_{\text{rev}}(A) = P_{\text{ave}}(A) \times \sum_{i=1}^{n} (w_i Q_i) \tag{4.10}$$

where $P_{\text{rev}}(A)$ represents the "installation-specific" probability or frequency of occurrence of the event A, $P_{\text{ave}}(A)$ signifies the industry-average probability of occurrence of the event A, w_i is the weight assigned to RIF No. i in relation to the event A, Q_i is a measure of the status of RIF No. i in relation to the event A, n is the number of RIFs.

Table 4.10 Description of RIFs

RIF group	RIF	Description
Personnel	Workload/stress	The overall workload experienced by individuals, encompassing all tasks and activities undertaken
	Competence	The individual's competence, experience, system knowledge and training relevant to their role
	Fatigue	The state of exhaustion experienced by an individual, often resulting from extensive use of time in activities
	Work environment	The physical conditions of the workplace, including factors such as noise, lighting, and vibration
Task	Methodology	The systematic approach employed to execute a particular task
	Task supervision	Oversight of specific tasks conducted by a designated supervisor
	Task complexity	The degree of complexity associated with a specific task
	Time pressure	The constraints imposed by time during the planning, execution and completion of a specific task
	Tools	The availability and operability of essential tools required for task execution
	Spares	The availability of spare parts necessary for task performance
Technical system	Equipment design	The design of equipment and systems, including aspects such as flange and valve types
	Material properties	The characteristics of selected materials concerning factors such as corrosion, erosion, and fatigue resistance
	Process complexity	The overall intricacy of the process plant
	Human machine Interface (HMI)	Aspects such as ergonomic considerations, equipment labeling, feedback mechanisms from valves, alarms, etc.
	Maintainability/ accessibility	The ease of maintenance of equipment and systems, including access to valves and flanges and the spatial requirements for tool usage
	System feedback	The mechanisms through which errors and failures are promptly detected
	Technical condition	The state of the technical system
Administrative control	Procedures	The quality and availability of established procedures and task descriptions

(continued)

Table 4.10 (continued)

RIF group	RIF	Description
	Work permit	The system governing work permits, including application, review, approval, follow-up, and control processes
	Disposable work descriptions	The quality and availability of disposable work descriptions, including safe job analysis, isolation plans, etc.
	Documentation	The quality, availability, and updating of technical documents, such as drawings, piping and instrumentation diagrams (P& IDs), etc.
Organizational factors	Programs	The scope and quality of programs for preventive maintenance (PM), condition monitoring, inspections, third-party work control, use of self-control or checklists, etc. A critical aspect is whether PM and condition monitoring programs are specified
	Work practice	The common practice employed during the execution of work activities, including adherence to procedures and checklists, acceptance of shortcuts and prioritization of time over quality
	Supervision	Oversight on the platform, including follow-up of activities, adherence to plans and compliance with deadlines
	Communication	The exchange of information among various stakeholders, including area platform managers, supervisors, area technicians, maintenance contractors and central control room (CCR) technicians
	Tidiness and cleaning	The general cleaning and tidiness of different areas within the platform
	Support systems	The quality of data support systems, such as information systems
	Acceptance criteria	The specific definitions of acceptance criteria related to condition monitoring, inspection, etc.
	Simultaneous activities	The number of concurrent activities, either planned or unplanned
	Management of changes	The processes involved in implementing changes and modifications

Thereinto, Q_i is determined in the following way:

$$Q_i(s) = \begin{cases} P_{\text{low}}/P_{\text{ave}} & \text{if } s = A \\ 1 & \text{if } s = C \\ P_{\text{high}}/P_{\text{ave}} & \text{if } s = F \end{cases} \tag{4.11}$$

Table 4.11 Scoring criterion for evaluating the status of RIFs

Score	Grade description
A	The status aligns with the highest standard in the industry
B	The status aligns with a level that exceeds the industry average
C	The status aligns with the industry average
D	The status aligns with a level that is somewhat below the industry average
E	The status aligns with a level that is significantly below the industry average
F	The status aligns with the lowest standard in the industry

where s is the score of RIF No. i, P_{low} and P_{high} are determined as the lower and upper limits for P_{ave} by expert judgment, respectively.

In order to assign values to Q_i in the case of $s = B$, it is assumed that a linear relationship exists between $Q_i(A)$ and $Q_i(C)$. In the same way, a linear relationship is also assumed when assigning values to Q_i for $s = D, E$.

Module 4 Calculation of the Installation-Specific Risk.

The risk associated with hydrocarbon leaks at a specific installation is determined by utilizing the installation-specific input probabilities and frequencies for all events within the risk model. The incorporation of these revised probabilities and frequencies yields an updated risk picture, which reflects the specific technical and non-technical RIFs pertinent to the installation.

In the year 2018, the authors enhanced the BORA method to more accurately reflect and analyze the impact of specific technical (hard) and non-technical (soft) RIFs on barrier performance. A new RIF-Index is developed to identify and categorize various RIFs associated with all basic events. Experts evaluate these RIFs using a pre-defined fuzzy scoring criterion. To determine the relative importance of the RIFs, a modified fuzzy analytic hierarchy process (F-AHP) that addresses fuzzy consistency is employed. Consequently, the industry-average frequencies and probabilities are updated through a comprehensive assessment, which integrates the priority weights and the status of the RIFs. Additional methodological details can be found in Zhen et al. (2018).

4.3.2 Risk Modelling: Integration of Organizational, Human and Technical Factors

The Risk_OMT method, as articulated by Vinnem et al. (2012) and Gran et al. (2012), emphasizes a more comprehensive modelling of RIFs and their impact on the operational barriers' performance. The Risk_OMT method signifies an advancement of the research conducted in the BORA and Operational Condition

Safety (referred to as OTS in Norwegian) projects, as detailed by Vinnem et al. (2007). The interconnection between these methods is depicted in Fig. 4.14.

In the framework of BORA modelling, the RIF structure is designed to be one-level, where all RIFs are assigned equal structural importance (Aven et al. 2006). The specific RIFs are selected from the general list, as has been presented in Table 4.10.

The OTS method builds upon the BORA method and focuses on the impact of HOFs on the barrier performance and the mechanism of operational barriers in reducing the risk of major accidents. The OTS method comprises seven performance standards: (1) work practice, (2) competence, (3) procedures and documentation, (4) communication, (5) workload and physical working environment, (6) management and (7) management of control.

The modelling principle of the Risk_OMT method is depicted in Fig. 4.15. The modelling process begins with a set of work operations. The various types of human failures that can lead to leak scenarios are defined as initiating events, such as incorrect blinding or isolation. Typically, one or more barriers are designed and implemented to prevent an initiating event from progressing into the leak. This is exemplified by the barriers labeled "End Control 1" and "End Control 2." The performance of the barrier systems and critical work tasks that contribute to the initiating events is modelled through the fault tree. The impact of HOFs on the basic events in the fault tree is modelled through RIFs. Ultimately, the interrelations between the RIFs are modelled by BNs. The choice of BNs can be attributed to their inherent flexibility and suitability to effectively integrate the impact of HOFs into risk modelling.

This section presents an overview of the Risk_OMT method, focusing on its fault tree model, the generic RIF model, and the implementation of BNs.

Fault Tree Model

The TOP event in the fault tree is defined as "failure or degradation of the work task or the barrier system". The causes of the TOP event are categorized into two groups, as illustrated in Fig. 4.15:

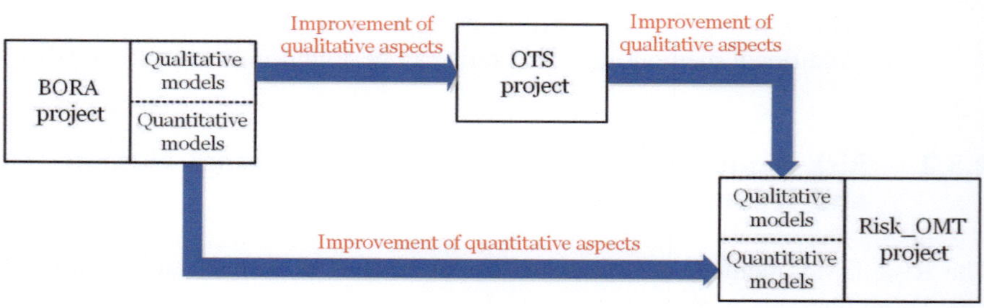

Fig. 4.14 The interconnection between BORA, OTS and Risk_OMT methods, adapted from Vinnem et al. (2012) with Elsevier permission

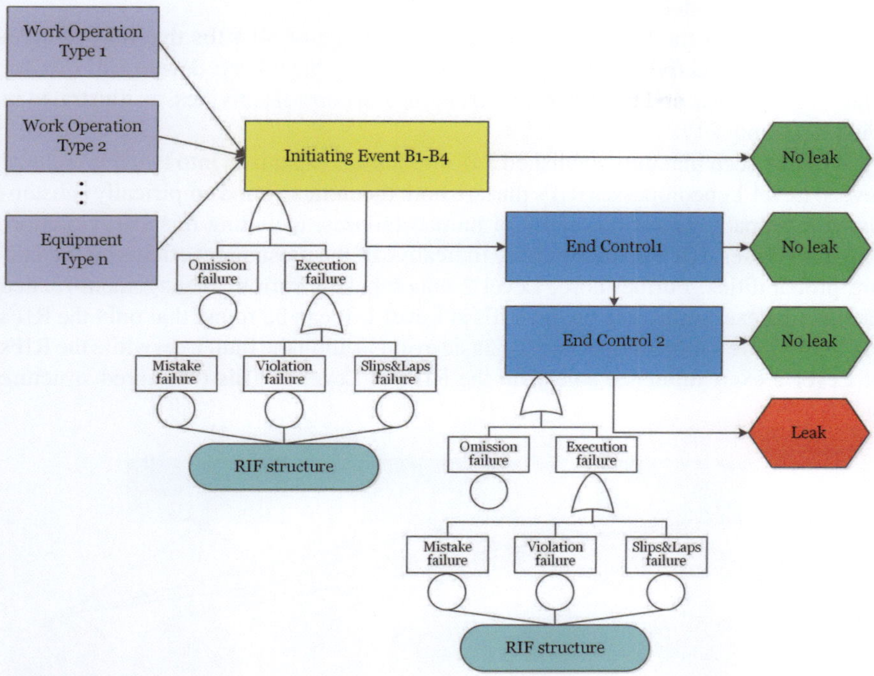

Fig. 4.15 The modelling principle of the Risk_OMT method, adapted from Vinnem et al. (2012) with Elsevier permission

1. Failure of omission, which refers to the absence of execution of the prescribed activity.
2. Failure of execution, which refers to inadequate actions that may lead to failures. This group includes several types of human failures, encompassing violations, human errors, and sabotage (Reason 1990):
   - Violation refers to deliberate (but not necessarily reprehensible) deviation from established practices that are deemed essential for the safe operation of potentially hazardous systems, as agreed by designers, managers, and regulatory agencies.
   - Human error can be categorized into two distinct types: mistakes, and slips and lapses. Mistakes arise from mis-interpretations of procedures, or failures in judgment and inferential reasoning related to the prescribed tasks. Slips and lapses refer to actions that represent unintended deviations from the established practices outlined in formal procedures.
   - Sabotage refers to deliberate actions aimed at damaging the system. It is classified as a marginal phenomenon and is excluded from consideration in the Risk_OMT method.

Generic RIF Model

In order to cover the broadest possible sample space of all RIFs that may affect a specific type of activities, two generic RIF models have been developed: one for planning activities and the other for execution and control activities, as illustrated in Figs. 4.16 and 4.17.

It can be seen that the established RIF structure is organized into two hierarchical levels. Level 1 encompasses RIFs that are both theoretically and empirically substantiated to impact the various types of human failures, including mistake, violation, slips and lapses. RIFs at this level are indicative of the most direct effects on the failure probabilities. Furthermore, Level 2 encompasses various management-related factors that exert influence on the RIFs at Level 1. It can be found that only the RIFs at Level 1 have a direct impact on the basic events within the fault tree, while the RIFs at Level 2 exert influence solely on the RIFs at Level 1. This two-tiered structure

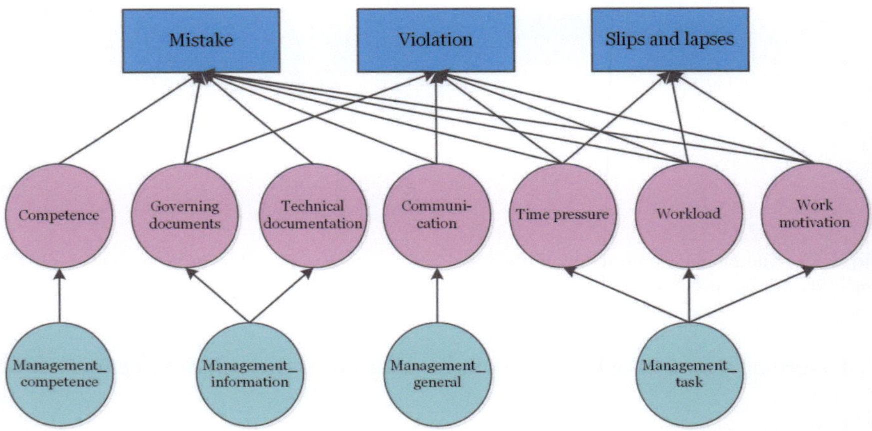

Fig. 4.16 The generic RIF model for planning activities, adapted from Vinnem et al. (2012) with Elsevier permission

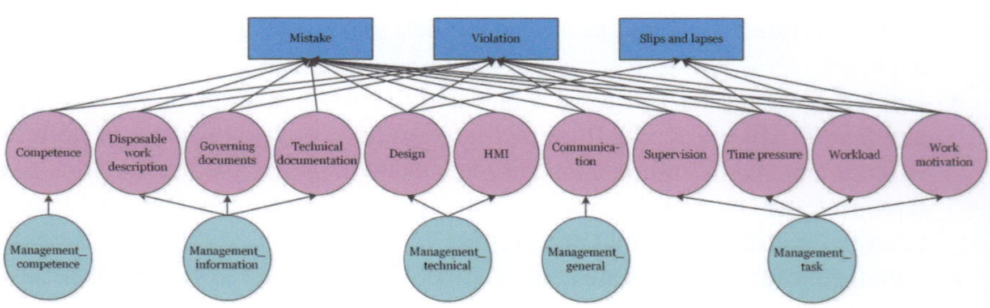

Fig. 4.17 The generic RIF model for execution and control activities, adapted from Vinnem et al. (2012) with Elsevier permission

serves as a mechanism to mitigate the uncertainty associated with the observations of RIFs at Level 1.

BN Implementation

The Risk_OMT model is constructed using BNs. Figure 4.18 illustrates a BN corresponding to the work task B1.B (Failure isolation/blinding). The upper section of the BN depicts the fault tree model, while the lower section represents the RIF model. The nodes within the fault tree are characterized by two states: failure and non-failure, whereas the remaining nodes are modelled with states A-F, as detailed in Table 4.11. Table 4.12 provides a summary of the interrelationships among the five types of nodes depicted in Fig. 4.18. It is noteworthy that a supporting level exists between the RIFs and the basic events, with these nodes summarizing the weights of the corresponding RIFs.

In the BN, AND-gates and OR-gates are represented through CPTs, which are constructed by inserting binary values (0-s and 1-s) in accordance with the structure of the fault tree. The CPTs for the basic events within the fault tree are formulated based on the human error probability (HEP) data collected from the BORA project.

The basic events are linked to the RIFs, which are evaluated using a scoring criterion ranging from A to F. This necessitates the collection of HEP data not merely as a singular or average probability, but also requires access to the spread

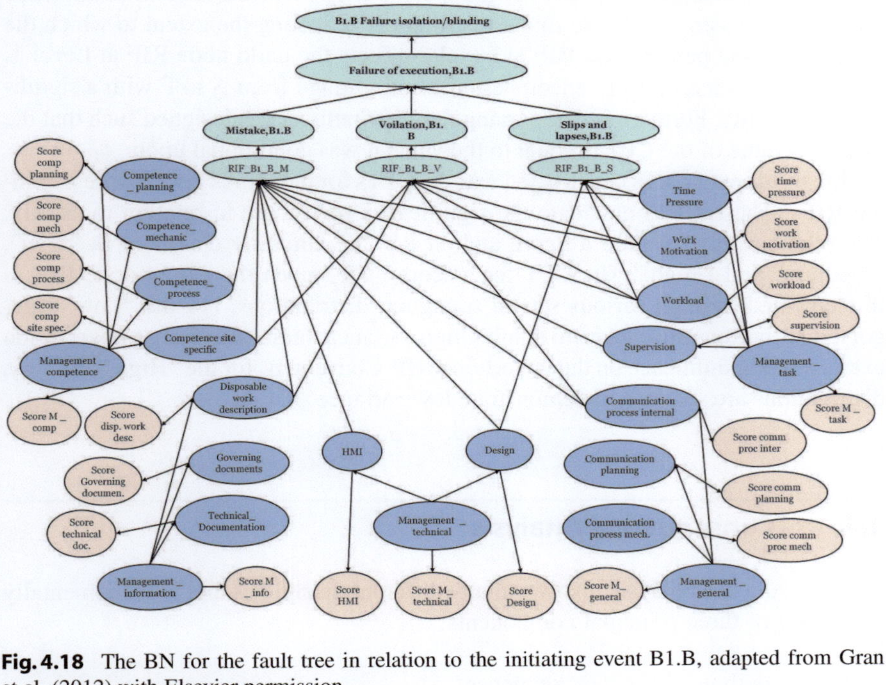

Fig. 4.18 The BN for the fault tree in relation to the initiating event B1.B, adapted from Gran et al. (2012) with Elsevier permission

Table 4.12 The relationship among nodes in a Bayesian network

Node	Parent node	Child node	State
Basic event	Supporting level RIF	Hazardous event	Failed/not failed
Supporting level RIF	Level 1 RIF	Basic event	A ~ F
Level 1 RIF	Level 2 RIF	• Supporting level RIF • Score	A ~ F
Level 2 RIF	–	• Level 1 RIF • Score	A ~ F
Score	• Level 1 RIF • Level 2 RIF	–	A ~ F

of this value. It proves most suitable to allocate the average HEP data to the score C, while the spread from the score C to other scores is represented through error fractions.

The importance of the RIFs, i.e., how strongly the RIFs affect the basic events, has been modelled through the application of designated weights.

The CPT for a RIF at Level 1, given its parent node (a RIF at Level 2), is developed according to the structural influence categorized into three levels: "Low", "Medium" and "High". The CPTs for RIF nodes are generated using distinct sets of triangular distributions. It is posited that each category of structural influence can be assigned the same relationship, irrespective of the specific RIFs involved. Structural influence serves as a measure of spread in the distribution, reflecting the extent to which the influence of the parent node RIF at Level 2 affects the child node RIF at Level 1. Each CPT is constructed to encompass the entire range from A to F with a significant probability. Furthermore, the triangular distributions are designed such that the expected value of the CPT is equal to the value it was conditional upon.

The third set of CPTs constructed was the CPTs for the scores given to the incoming RIFs. These CPTs elucidate the significance of a score in relation to an RIF. The relationships between a Score and an RIF are similarly classified as "Low", "Medium" and "High." The CPT for a score node, given the parent node RIF, is implemented through various sets of triangular distributions. For the "Low" category, distributions are chosen to exhibit high variance, indicating that an observation exerts minimal influence on the underlying RIF. Conversely, for the "High" category, distributions are selected to demonstrate low variance.

4.4 Consequence Analysis

The consequence analysis involves many different disciplines and is fundamentally composed of three principal components:

- The modelling of accident sequences.

- The physical consequence analysis, which involves the analysis of:
  - accidental loads associated with fire, explosion, and impact;
  - the response of systems and equipment when subjected to accidental loads;
  - escalation pertaining to the propagation path of accidents from the initial equipment to surrounding equipment and areas.
- The quantification of the ultimate consequences for personnel, the environment, and economic assets.

Notably, the greatest challenge of the consequence analysis lies in the modelling of accident sequences, corresponding to the steps taken as well as the timing elements associated with these sequences.

There are several quantitative techniques for modelling of accident sequences. The following presents an introduction to the most frequently employed techniques, specifically ETA and ESD. Ultimately, the topic of escalation analysis is also addressed.

The latest edition of NORSOK Z-013 standard (Standards Norway 2023) describes methods for risk analysis for typical major accident hazards to provide more standardization between analyses. This was not covered in the previous edition (Standards Norway 2010). These descriptions include detailed methods for calculation of leak duration, dispersion, ignition and fatality fractions, impairment of main safety functions as well as a detailed method for addressing Worst Credible Process Fire.

4.4.1 Event Tree Analysis

ETA is an inductive and visual methodology utilized to model the range of potential outcomes arising from a specific initiating event. Figure 4.19 provides a representative example of an event tree structure. The possible outcomes are determined by responding to a series of binary questions, typically framed as "yes" or "no". These questions frequently pertain to whether the barriers are functioning or not. Consequently, the ETA technique reflects the designers' way of thinking. A more comprehensive introduction to ETA can be found in the International Standard IEC 62502 (International Electrotechnical Commission 2010).

ETA is conducted with several key objectives, including:

- providing a qualitative description of the accident scenarios;
- determining the quantitative frequencies of the consequence classes associated with each accident scenario;
- evaluating the impact of key parameters through sensitivity analysis;
- identifying the significant contributors to each consequence class;
- recommending measures for risk reduction.

On the basis of IEC 62502 (2010), the procedure for ETA primarily encompasses six steps:

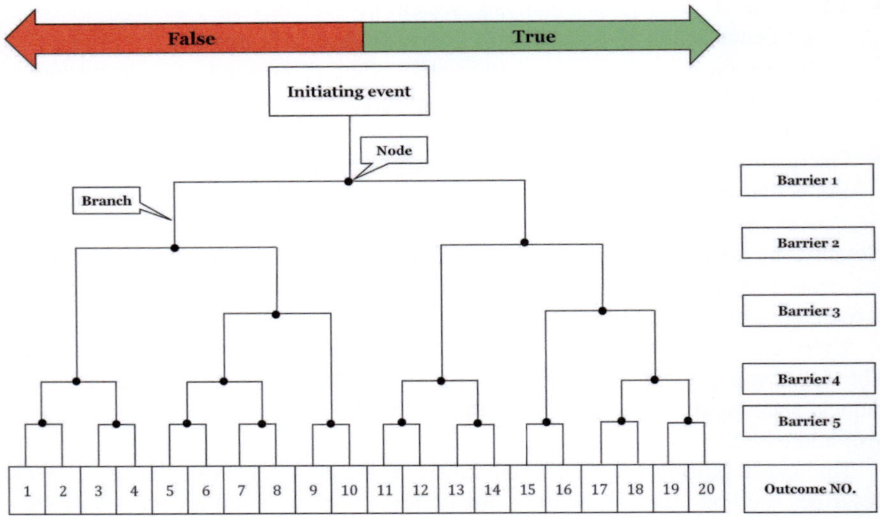

Fig. 4.19 A graphical representation of an event tree

STEP 1: Definition of the System or Activity.
This initial step involves defining the boundaries of the system or activity to be analyzed. The analytical boundaries are determined so as to restrict the detail level of the analysis as well as exclude specific types of events or conditions from consideration.

STEP 2: Identification of the Initiating Events.
This step identifies the hazards and potential initiating events of primary concern that will be addressed in the ETA. For an overview of hazard identification techniques, please refer to Sect. 4.1 of this text.

STEP 3: Identification of the Barriers and Physical Phenomena.
It is essential to identify the various barriers that may affect the progression from the initiating event to its final outcomes. Additionally, the physical phenomena, which describe evolution taking place inside and outside the system, should also be identified, such as ignition or meteorological conditions that will affect the progression and the final outcome of the initiating event.
These identified barriers and physical phenomena should be modelled to construct the event tree.

STEP 4: Definition of Accident Scenarios, and Construction of the Event Tree.
For each initiating event, the event sequences and outcomes will be defined.
The methodology for constructing an event tree is advised as follows:
(a) Position the initiating event at the uppermost side of the event tree;

(b) Position the barriers and physical phenomena along the right of the tree, for instance, in the chronological sequence in which they will influence the progression of the accident;
(c) At each node, identify the success (typically represented by a rightward branch) and failure (represented by a leftward branch) of each barrier.

It should be noted that too large event trees should be avoided.

STEP 5: Quantitative Analysis.
The quantitative analysis is conducted based on the constructed event tree, encompassing two primary aspects:

- Independent sequence of events. When all the conditional probabilities associated with the success or failure of barriers are independent of one another, the frequencies or probabilities of the outcomes stemming from the initiating event can be calculated. The calculation procedure is illustrated in Fig. 4.20.
- Fault tree linking. Fault trees could serve as a tool for determining the conditional probability of the barrier failure, as displayed in Fig. 4.21.

STEP 6: Reporting on the ETA Results.
The various outcomes are then analyzed with respect to their consequences and their impact on the ETA results. The qualitative and quantitative findings derived from the ETA are then converted into necessary actions.

Notably, two critical dimensions of dependencies ought to be highlighted:

(1) Functional dependencies, which necessitate clarity regarding whether the successful implementation of one barrier is contingent upon the successful implementation of another.
(2) Structural or physical dependencies, which typically lead to common cause failures that can result in multiple events. In particular, the common systems or functions that influence the various barriers should be identified.

4.4.2 Event Sequence Diagram

ESD serves as a graphical representation of an accident scenario, illustrating various pathways that culminate in distinct end states. Figure 4.22 presents a typical structure of an ESD alongside its symbolic representations. It is evident that an ESD functions fundamentally as a flowchart, with each pathway denoting a specific accident scenario that leads to distinct outcomes.

It is worth noting that ESD can also be transformed into an event tree to facilitate quantifying the accident scenarios. Nonetheless, it is thought that ESD provides a superior advantage compared to ETA in improving communication

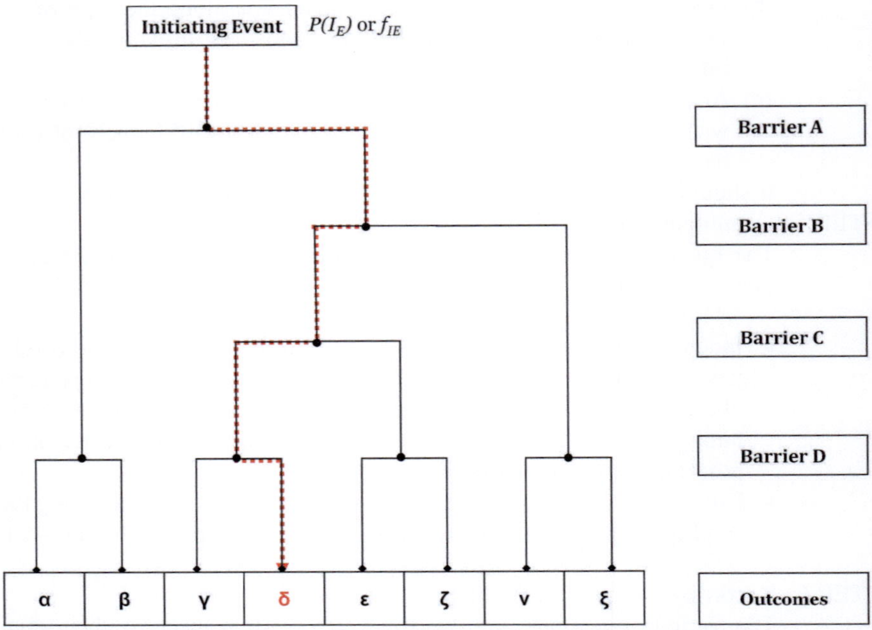

$P(\delta) = P(I_E \times A.\bar{B}.\bar{C}.D) = P(I_E) \times P(A|I_E) \times P(\bar{B}|I_E.A) \times P(\bar{C}|I_E.A.\bar{B}) \times P(D|I_E.A.\bar{B}.\bar{C})$

where, $P(A|I_E)$ denotes the probability of success of barrier A given the initiating event has occurred.

Fig. 4.20 The basic principle of calculating the frequency or probability of the outcome δ

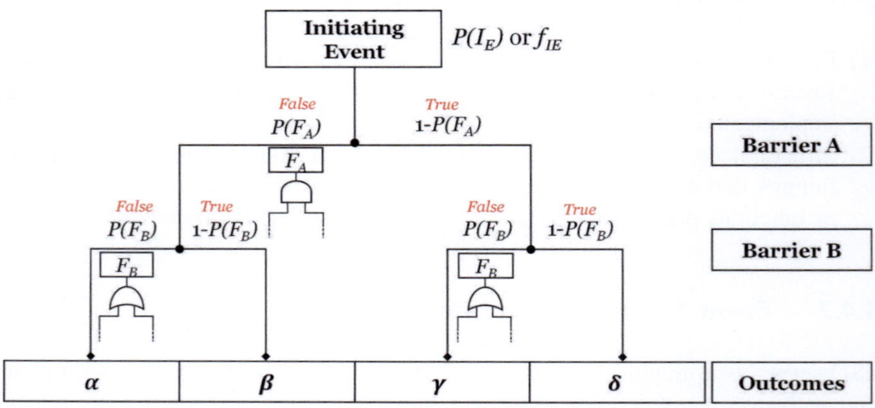

Fig. 4.21 Fault tree linking

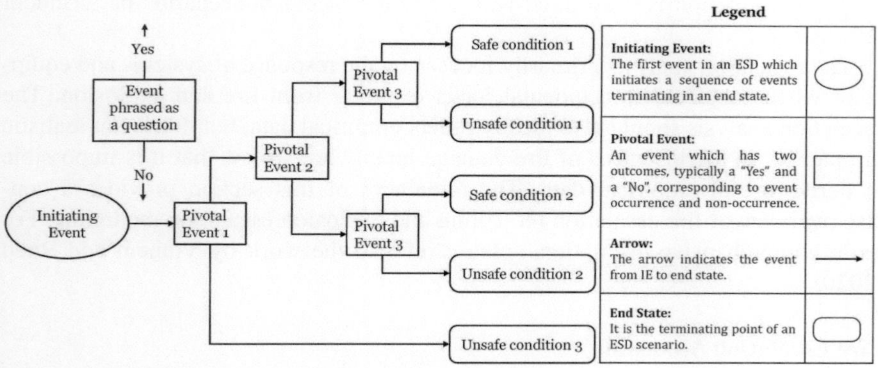

Fig. 4.22 A typical structure of ESD

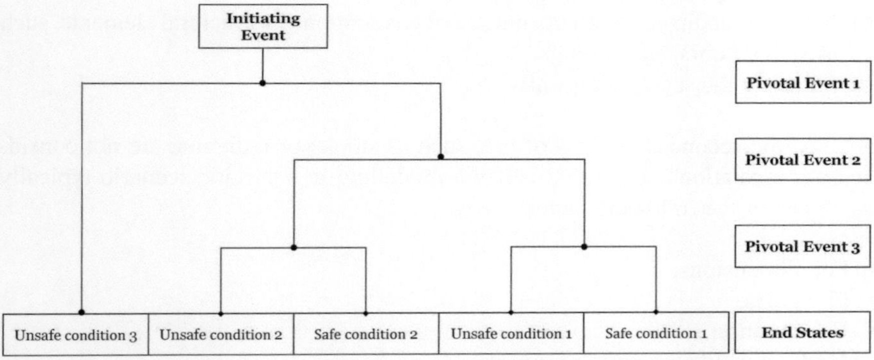

Fig. 4.23 Event tree conversion of the ESD presented in Fig. 4.22

between risk analysts and other relevant stakeholders. Figure 4.23 demonstrates the conversion principle from ESD to ETA on the basis of the typical structure of ESD in Fig. 4.22. A more comprehensive discussion of ESD is available in the works of Swaminathan and Smidts (1999a, b).

4.4.3 Escalation Modelling

The concept of "escalation" can be interpreted in either the broad context or the narrow context. The broad interpretation encompasses the whole process, extending from the initiating event to the end events. Conversely, the narrow interpretation defines escalation as the secondary failure of containment resulting from the accidental effects. This book takes the narrow view of "escalation," while the broad interpretation is substituted with the terms “accident scenario” or “accident sequence”.

The escalation analysis primarily focuses on the response of systems and equipment when subjected to accidental loads resulting from fire and explosion. The escalation analysis should also ideally reflect empirical data, but this is not realistic in practice, as the instances of fire damage luckily are so few that it is impossible to derive useful empirical data. The remainder of this section provides a concise overview of fire escalation modelling and explosion escalation modelling. For more comprehensive discussions, please refer to the work by Vinnem and Røed (2020).

Fire Escalation Modelling

Fire escalation is essential for forecasting the potential spread of a fire beyond its initial location. The mechanisms of fire escalation primarily include:

(a) thermal impact from external flames;
(b) flames spreading through openings and penetrations in structural elements, such as walls, floors, and ceilings;
(c) failure of the segregating walls.

Notably, the secondary effects of fire, such as smoke or radiation, are not considered as “escalation”. The fire escalation modelling in a specific scenario typically encompasses the following factors:

- Fire dimensions,
- Fire types,
- Fire duration,
- Medium inside the pipes or vessels,
- Effect of fire protection barriers,
- Structural response to fire loads.

Explosion Escalation Modelling

Explosions have generated significant concern due to their high blast loads, which may lead to escalation in various ways:

(a) overall structural collapse,
(b) rupture and excessive deformation of explosion protection barriers,
(c) excessive deformation of decks, walls and process equipment, resulting in loss of containment in other areas,
(d) impairment of safety systems that results in their inability to function.

The escalation modelling of explosions ought to accurately represent the actual explosion loads as well as the structural and equipment capacities of the installation. The current practices suggest that the computational fluid dynamics (CFD) and structural analysis software should be employed for explosion analysis. Particularly, the response analysis should account for elastic and plastic structural response, as well as the influence of large deformation.

4.5 Online Risk Analysis

Havtil has consistently asserted that the conventional risk assessment techniques, which are employed within the Norwegian petroleum sector, fall short in effectively supporting operational decisions. In response to the regulatory authority's concern seriously, the concept of online risk management was initially introduced by the author (Vinnem et al. 2015), with the objective of providing operators a real-time risk picture along with pre-warnings regarding potential deviations. Building upon this foundation, the authors (Zhen et al. 2020) have further developed an online risk monitoring and decision support (ORMDS) framework, as illustrated in Fig. 4.24. This framework consists of four crucial modules:

(a) Data collection,
(b) Online risk models (ORMs),
(c) Autonomous assistant system (AAS),
(d) Organizational management system (OMS).

This section utilizes the ORMDS framework as a case study to explore the authors' insights and interpretations regarding online risk analysis.

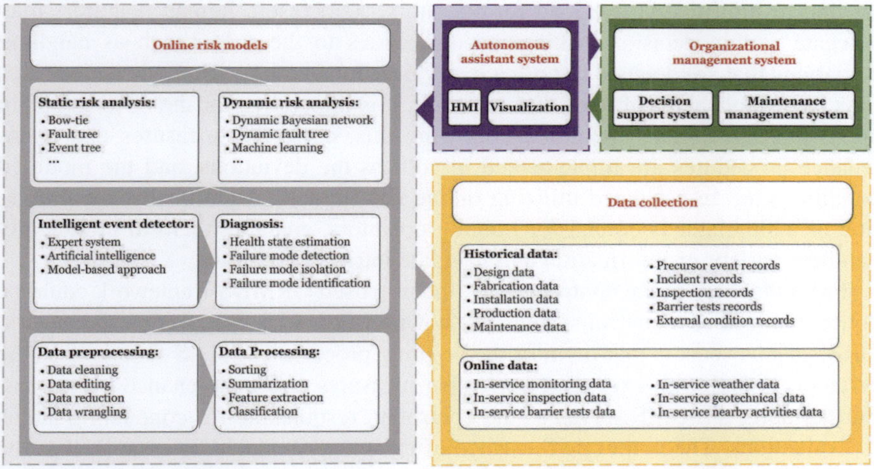

Fig. 4.24 The framework for online risk management system, reprinted from Zhen et al. (2020) with Elsevier permission

First of all, the online data, along with historical data, constitutes the foundation for the ORMs development. The online database is specifically designed to collect the condition monitoring data in accordance with real-time monitoring and data transmission techniques, facilitating an understanding of the system's operational status. Based on the collected data, potential deviations could be identified according to the data processing, intelligent event detection, as well as diagnostic techniques. Recent advancements in communication technologies, such as optical fiber sensing and satellite communication, are expected to establish a robust basis for online connectivity, particularly with the potential applications of 5G technology. As a result, the datasets obtained from a multitude of sources are expected to expand rapidly, resulting in a volume of big data that surpasses the processing capabilities of conventional tools within acceptable timeframes. Consequently, data pre-processing and processing techniques play a crucial role in ensuring data quality, consistency and suitability. Subsequently, the refined data are analyzed by an intelligent event detector, which works alongside the expert system and AI (Artificial Intelligence) model. This advanced detector facilitates the diagnostic process for identifying and clarifying any deviations.

The ORMs are able to be constructed by utilizing the static and dynamic risk analysis approaches, including but not limited to the bow-tie model, FTA, ETA, DBN, and machine learning techniques. The results derived from these ORMs are displayed through visualization tools integrated within an AAS. When the analyzed risk associated with a diagnosed deviation exceeds the established warning threshold, the emergency response is swiftly initiated to commence the coordinated decision-execution workflow. Notably, large and complex offshore production systems necessitate the management of extensive information and the execution of numerous actions to address deviations that pose major accident risks, potentially leading to information overload for operators. As a result, human operators may delegate certain decision-making responsibilities to the AAS, such as handling deviations that are deemed to pose a lower risk. In situations where deviations surpass the established authority of the system, it is imperative for the AAS to obtain approval from human operators. Following this, corrective measures and maintenance procedures are implemented to address the deviations, and the modified conditions are re-evaluated utilizing online risk models. Additionally, the utilization of a simulator is crucial for training operators, enabling them to proficiently handle a variety of information and navigate intricate scenarios.

From the view of the control theory, the proposed ORMDS framework could be conceptualized as a control system, as shown in Fig. 4.25. Within this framework, ORMs act as sensors that monitor the ongoing process. The AAS functions as the actuator, tasked with executing corrective measures and maintenance operations. Furthermore, the OMS serves as the controller, responsible for conducting audits or evaluations. The output performance of the actuator should be compared to the established performance metrics, with feedback sent back to the actuator thereafter. Effective communication is essential for the seamless operation of these modules.

The proposed ORMDS framework underscores the importance of investigating the potential for optimizing the use of current sensors, instrumentation and AAS

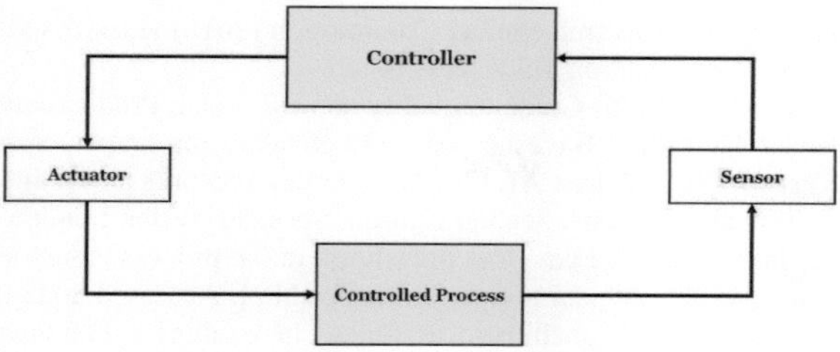

Fig. 4.25 The principle of a control system

to enhance the online decision-making support and the development of innovative monitoring solutions. As a result, an independent ORMDS framework could be established for offshore petroleum installations, serving as an additional layer of protection. Notably, the AAS's advisory capabilities will enhance the ability of operators to make well-informed and prompt decisions.

4.6 Questions

Q1: Please list three methods utilized for causal analysis and consequence analysis.
Q2: Please explain how PHA, FTA and ETA could be used in a risk analysis.
Q3: In the PHA and FMECA, what is used for risk ranking?
Q4: What could be the limitation of using Bayesian networks in a risk analysis?
Q5: Please prepare a PHA for a collision scenario between FPSO and shuttle tanker.

References

Aslansefat K, Kabir S, Gheraibia Y, Papadopoulos Y (2020) Dynamic fault tree analysis: state-of-the-art in modelling, analysis and tools. In: Reliability management and engineering: challenges and future trends. CRC Press, pp 73–112

Aven T, Sklet S, Vinnem JE (2006) Barrier and operational risk analysis of hydrocarbon releases (BORA-release). Part I. Method description. J Hazard Mater 137:681–691

BayesFusion (2025) Available: https://www.bayesfusion.com/genie/. Accessed 2025/03/23

Crawley F, Tyler B (2015) HAZOP: guide to best practice for the process and chemical industries, 3rd edn. Elsevier Ltd

Gran BA, Bye R, Nyheim O, Okstad E, Seljelid J, Sklet S, Vatn J, Vinnem JE (2012) Evaluation of the risk OMT model for maintenance work on major offshore process equipment. J Loss Prev Process Ind 25:582–593

Hugin Expert (2025) https://www.hugin.com/. Accessed 2025/03/23

International Electrotechnical Commission (2006a) Analysis techniques for system reliability-procedure for failure mode and effects analysis. IEC 60812

International Electrotechnical Commission (2006b) Fault tree analysis (FTA). IEC 61025

International Electrotechnical Commission (2010) Analysis techniques for dependability event tree analysis (ETA). IEC 62502

International Electrotechnical Commission (2016) Hazard and operability studies (HAZOP studies)—application guide. IEC 61882
Ishikawa K (1976) Guide to quality control. Asian Productivity Organization, Tokyo
Jensen FV (2001) Bayesian networks and decision graphs. Springer, London
Kjærulff UB, Madsen AL (2014) Bayesian networks and influence diagrams: a guide to construction and analysis. Second Edition, Springer-Verlag London Ltd
Mannan S (2012) Lees' loss prevention in the process industries, 4th edn. Elsevier Inc
Norsys (2025) https://norsys.com/netica.html. Accessed 2025/03/23
Pearl J (2001) Probabilistic reasoning in intelligent systems: networks of plausible inference. Morgan Kaufman, San Mateo
Petri CA (1962) Kommunikation mit Automaten. University of Bonn, Bonn, Germany
Reason J (1990) Human error. Cambridge University Press, Cambridge
Ross SM (1996) Stochastic processes. Wiley, New York
Skogdalen JE, Vinnem JE (2011) Quantitative risk analysis offshore-human and organizational factors. Reliab Eng Syst Saf 96:468–479
Standards Norway (2010) Risk and emergency preparedness assessment. NORSOK Z-013. Rev.3
Standards Norway (2023) Risk and emergency preparedness assessment. NORSOK Z-013. Rev.4
Swaminathan S, Smidts C (1999a) The event sequence diagram framework for dynamic probabilistic risk assessment. Reliab Eng Syst Saf 63:73–90
Swaminathan S, Smidts C (1999b) The mathematical formulation for the event sequence diagram framework. Reliab Eng Syst Saf 65:103–118
Vinnem JE (2007) Operational risk analysis—total analysis of physical and non-physical barriers. Generalization report. Report No. 200254-07
Vinnem JE, Røed S (2020) Offshore risk assessment: principles, modelling, and applications, 4th edn. Springer-Verlag, London Ltd
Vinnem JE, Haugen S, Kongsvik T, Seljelid J, Steen S, Sklet S, Thomasssen O (2007) Operational safety condition—concept development. In: Proceedings of ESREL 2007, Stavanger
Vinnem JE, Bye R, Gran BA, Kongsvik T, Nyheim OM, Okstad EH, Seljelid J, Vatn J (2012) Risk modelling of maintenance work on major process equipment on offshore petroleum installations. J Loss Prev Process Ind 25:274–292
Vinnem JE, Utne IB, Schjølberg I (2015) On the need for online decision support in FPSO-shuttle tanker collision risk reduction. Ocean Eng 101:109–117
Zhen XW, Vinnem JE, Peng CY, Huang Y (2018) Quantitative risk modelling of maintenance work on major offshore process equipment. J Loss Prev Process Ind 56:430–443
Zhen XW, Vinnem JE, Han Y, Peng CY, Yang X, Huang Y (2020) New risk control mechanism for innovative deepwater artificial seabed system through online risk monitoring system. Appl Ocean Res 95:102054

5 Fifth Lesson: Principle of Risk-Based Design

Risk-based thinking ensures these risks are identified, considered and controlled throughout the design and use of the quality management system.

—International Organization for Standardization (2015)

Abstract

The risk-based design approach represents a transformative shift from prescriptive design rules to goal-based solutions. The offshore petroleum enterprises have expressed their enthusiasm for supporting this promising design approach. This lesson starts with exploring the need and challenges underlying the implementation of risk-based design. Then, the lesson outlines regulatory requirements for risk-based design and clarifies its relationship with risk modelling. A focus is placed on the safety integrity level (SIL), which serves as a measure of the level of risk reduction provided by a safety function or the reliability of a safety system. Finally, practical considerations for risk-based design in the context of accidental loads are discussed. This lesson aims to equip readers with a comprehensive understanding of risk-based design principles and their practical applications.

5.1 Needs and Challenges of Risk-Based Design

In recent years, risk-based design has been receiving growing attention within the offshore petroleum sector. By systematically integrating risk assessment into the design process, the enterprises can proactively address hazards and risks while also optimizing the allocation of available resources. This section focuses on the critical examination of the need and challenges in implementing risk-based design.

X. Zhen and J.-E. Vinnem, *Ten Lessons on Risks*,
https://doi.org/10.1007/978-3-031-99064-9_5

5.1.1 The Needs of Risk-Based Design

The conventional prescriptive design approach relies heavily on compliance with regulatory requirements and technical standards. It can be envisaged that the prescriptive design approach fails to prioritize dynamic and context-sensitive solutions tailored to the hazards and risks. In this context, risk-based design emerged as a response to these challenges, and the design paradigm began to shift from prescriptive rules to goal-based solutions. Risk-based design emphasizes the selection and development of preventive and protective measures based on the specific hazards and risks that have been identified. Clearly, this design paradigm integrates a substantial utilization of risk assessment into the conventional design process, enabling optimization of solutions tailored to each specific case. In addition, it can be noted that the term "risk-informed design" is also frequently used to emphasize that decisions should not merely take the risk assessment results into account. Nonetheless, their distinction is challenging to tell apart in practice. This book primarily concentrates on "risk-based design".

Up to the present, though the offshore petroleum enterprises have shown their enthusiasm in supporting the risk-based design approach, it is noteworthy that it has not been widely adopted within the offshore petroleum sector. The following questions could underscore the crucial implications of this critical discrepancy.

- Are the design solutions based on risk analysis perceived to be trustworthy?
- Is risk analysis perceived to be trustworthy as a tool?
- Are the risk assessors perceived to be trustworthy?
- Is the quality of data perceived to be trustworthy?

The above questions will be discussed in the subsequent sections, focusing on the primary types of risk associated with fire, explosion, collision impact and falling loads impact.

The need for a risk-based design approach can be further exemplified by design against blast loads. Prior to 1990, the design approach was characterized by:

- defining worst-case conditions for explosion;
- simulating or calculating the blast loads for the worst-case conditions;
- establishing and implementing design solutions for the worst-case conditions.

Nonetheless, towards the end of the 1990s, the situation with respect to design against blast loads underwent a significant transformation due to the findings from the BFETS (Blast and Fire Engineering for Topside Structures) test project, as follows:

- Design for worst-case conditions was either unfeasible or economically prohibitive;
- An alternative approach was needed;
- The probabilistic approach became an industry standard.

Table 5.1 The risk-based design scope

Structural system	Passive safety system	Active safety and emergency system
• Support structure • Hull structure • Deck structure • Buoyancy compartment • Subsea installations	• Passive fire protection • Blast protection • Dropped load protection of topside systems	• Gas detection • Fire detection • Emergency shutdown valves • Process safety valves • Process safety instrumentation • Fire water supply • Deluge systems • Sprinkler systems • Ventilation system

- The probabilistic modelling approach of blast loads was developed.

Consequently, for explosion scenarios, adopting a probabilistic approach is no longer optional but a necessity driven by technical and economic constraints.

Table 5.1 presents the systems on offshore petroleum installations that may need to be considered in risk-based design. Notably, the design of both the structural system and the passive safety system can be classified as part of structural design.

5.1.2 The Challenges of Risk-Based Design

The challenges of risk-based design could be attributed to many factors, such as the increased costs of typical solutions, uncertainties with respect to analytical results, etc. In addition, the subject of "risk-based design" itself is controversial and challenging due to issues of trustworthiness.

For standard installations, the challenges may correlate with the importance of design accidental loads and premises, which hold particular significance for passive structural fire protection, design to blast loads, as well as dimensioning against impact from collision and falling objects, as outlined in Table 5.2.

Notably, all the design aspects, which are of importance to the standard installations, are true of the minimum ("normally unmanned") installations, especially during periods when the installation is manned for maintenance. The following design aspects may also be crucial for minimum installations:

- The extent of utilization of active safety and emergency systems;
- The availability requirements for active safety and emergency systems;
- The protection against external impact for minimum structures.

What calls for special attention is that some new hazards have been introduced at the border area between operation and design. A pertinent example of this challenge could be illustrated by the capsizing accident of the semi-submersible

Table 5.2 Design to accidental loads encountering challenges

Design for accidental loads	Structural systems and equipment
• Passive structural fire protection	• Fire barriers • Fire resistance of equipment • Fire resistance of structures
• Design to blast loads	• Blast and fire barriers • Blast resistance of equipment • Blast resistance of structures
• Dimensioning against impact from collision	• Steel platforms • Floating concrete structures • Floating installations, if the deck structures couldn't provide reserve buoyancy when damaged
• Dimensioning against impact from falling objects	• Impact resistance of buoyancy elements • Tension leg platform (TLP), if not designed to handle the dynamic failure of a single tether or the flooding of a damaged compartment

platform Ocean Ranger (Royal Commission on the "Ocean Ranger" Marine Disaster 1984), where ballast pumps were designed at only one end of the pontoons and thus certain tanks cannot be emptied under certain dangerous conditions of inclination.

5.2 Regulation Requirements

The Norwegian regulations include provisions of risk-based design on structural systems and equipment, and the NORSOK Standard Z-013 (Standards Norway 2010, 2023) presents two pertinent definitions.

Dimensioning accidental load (DiAL)	Accidental load that a function or a system shall be able to withstand for a given period of time to meet the defined acceptance criteria for risk and minimum requirements in the regulations
Dimensioning accidental event (DiAE)	Accidental events that serve as the basis for layout, dimensioning and use of installations and the activity at large

Notably, the DiAL is commonly defined on the basis of the DiAE. Typically, the DiAL is derived from a risk assessment and is defined as the load with an annual occurrence probability of 1×10^{-4}. The following cases could be considered for the use of DiAL.

- Design of passive fire protection,
- Design of passive blast protection,
- Design of fire divisions,

- Combat of acute environmental pollution,
- Selection of survival suits for personnel protection in an emergency,
- Specification of emergency units for personnel treatment in an emergency.

It is crucial to emphasize the minimum or deterministic stipulation, such as the capacity requirement of free-fall lifeboats, supply requirement of fire water, etc., which cannot be changed by any kind of analysis.

The NORSOK Standard N-001 (Standards Norway 2012) asks for the accidental limit states (ALS) check for structural design. This check ensures that accidental actions do not result in a complete loss of integrity or performance of the structure and related maritime systems. The ALS check consists of two steps:

STEP 1: Resistance to accidental actions.
The structure and related maritime systems should be checked to maintain the prescribed load-carrying function under the defined accidental actions.

STEP 2: Resistance in damaged condition.
Following local damage that may have been demonstrated under the first step or following more specifically defined local damage, the facility shall continue to resist defined environmental conditions without suffering extensive failure, free drifting, capsizing, sinking or extensive damage to the external environment.

It is evident that the ALS check is intrinsically linked to risk analysis for the purpose of determining what accidental actions.

5.3 Relationship with Risk Modelling

Although the close relationship between risk-based design and risk analysis is obvious, the implementation of risk analysis within the framework of risk-based design encounters significant challenges. To address these challenges, the following five questions will be examined and clarified in this section:

Q1: What is the most appropriate type of risk analysis?
Q2: What selection criteria should be employed for the DiAE in an event tree?
Q3: What crucial elements should be modelled in an event tree?
Q4: Which type of consequence models should be employed in the risk analysis?
Q5: Is the risk analysis model sensitive to changes in active safety systems?

5.3.1 Suitability of Risk Analysis

The suitability of risk analysis involves both the type of risk analysis and RTC, and is significantly influenced by the type of systems and equipment to be analyzed. The following discussions are classified as structural and passive safety systems, active safety and emergency systems.

Structural and Passive Safety Systems

Under Norwegian legislation, the structural systems must be designed according to the prescribed approach outlined in the relevant regulations, which also define the acceptance limits of structural consequences and exceedance frequency. This indicates that the extensive analysis of consequences is not necessary and thus the uncertainties are rather limited. Consequently, the structural systems can be analyzed individually.

The definitions of the most severe consequences and the occurrence frequency for the structural systems are presented explicitly:

• Consequence	Functional capability of the structural system in the damaged condition
• Frequency	Occurrence frequency for accidental loads on the structural system

The design of passive safety systems is coupled with the structural systems to some degree as their dimensioning aspects correlate with the load specifications (e.g., the thermal load capacity for fire protection, the blast load capacity for explosion protection, and the impact capacity for dropped object protection). Therefore, the type of risk analysis and RTC adopted for the design of passive safety systems ought to be closely aligned with those of the structural systems. It is noteworthy that the Norwegian facilities regulations (Havtil 2020) have established requirements for escalation prevention, which are similar to the requirements on the design of structural systems.

Active Safety and Emergency Systems

With respect to risk-based design, it is most challenging to handle the active safety and emergency systems due to the diverse nature of their design parameters, as shown in Table 5.3. The diversity of these parameters underscores the importance of determining which parameters are suitable for risk-based design of active safety and emergency systems. Especially, it is noteworthy that the current risk analysis approaches are difficult to effectively model certain parameters, such as valve types and the locations of various equipment.

In most cases, the analysis results related to the active safety and emergency systems serve to determine design premises rather than design loads.

The types of risk analysis and RTC are suitable for the design of active safety and emergency systems, provided that the results of risk analysis can be categorized into different scenarios and severity levels.

Table 5.3 An overview of the parameters regarding the design of active safety and emergency systems

System	Parameters
• Gas detection	• Type of detectors • Location of detectors • Detection logic • Detection availability • Test interval
• Fire detection	• Type of detectors • Location of detectors • Detection logic • Detection availability • Test interval
• Emergency shutdown valves	• Location of valves • Sectionalization • Type of valves • Closing time • Availability • Fire protection
• Fire water supply	• Number of pumps • Type of pumps • Reliability of power supply • Availability • Independence from operational systems
• Deluge system	• Location of valves • Sectionalization • Location of deluge heads • Availability • Test interval
• Sprinkler system	• Location of valves • Sectionalization • Location of deluge heads • Availability
• Ventilation system	• Natural/mechanical systems • Location of ducts • Separate systems • Availability
• Fire water distribution	• Routing of supply lines • Redundancy • Sectionalization • Inspection and test intervals

(continued)

Table 5.3 (continued)

System	Parameters
• Process safety valves	• Location of valves • Sectionalization • Type of valves • Closing time • Depressurization time • Fire protection
• Process safety instrumentation	• Need for additional instruments

5.3.2 Use of Event Trees

Design of Active Safety and Emergency Systems

The event tree analysis is primarily used in the risk-based design of active safety and emergency systems. In this process, it is required to construct a sophisticated event tree so as to determine the design premises through risk analysis.

When conducting a risk analysis study for the design of active safety and emergency systems, several critical factors require meticulous evaluation:

- Release rate.
- Geometry of gas leaks.
- Size of flammable clouds.
- Performance of drainage systems.
- Overpressure from gas explosion.
- Impact of smoke.
- Potential sources of ignition.
- Fire sizes and the radiation impact.
- Differentiation between ignitions that may result in a fire versus explosion.

Design of Structural and Passive Safety Systems

In the risk-based design of structural and passive safety systems, the selection of event tree types (whether coarse or sophisticated) should be aligned with the corresponding type of consequence models. The sophisticated event trees should be employed when the sophisticated consequence models are required, and vice versa.

5.3.3 Use of Consequence Models

In the risk-based design of structural and passive safety systems for accidental load resistance, sophisticated consequence models should be chosen. Table 5.4 presents two options for the use of the sophisticated consequence models.

Due to the practical constraints on the modelling of complex physical processes, such as fire loads, the second option is preferred for the following advantages:

Table 5.4 Options on the use of sophisticated consequence models

No.	When	Implications
1.	Use of sophisticated consequence models in risk analysis	• The integration of sophisticated consequence models into the risk analysis process • The direct extraction of DiALs from risk analysis results
2.	Use of sophisticated consequence models during the design phase	• The risk analysis is conducted with empirical consequence models, from which preliminary DiALs and DiAEs are derived • Advanced consequence modelling is subsequently performed to confirm or modify the preliminary DiALs

- The risk analysis utilizing empirical consequence models can be conducted quickly.
- The expenditure on extensive resources and time can be minimized.
- Advanced consequence modelling can be restricted to the scenarios that have been selected as DiAEs.

The coarse modelling approach provides an additional advantage of being normally conservative, thereby enhancing the robustness of the design. What needs to be pointed out is that, while some years ago it was reasonable to use simplified models in some cases, the computers are now so powerful that sophisticated software can always be used.

However, even though computers are becoming increasingly faster, it is not feasible to model all combinations of parameters with CFD simulations. Thus, the state-of-the-art approach is to model a substantial set of representative scenarios with CFD simulations, and then integrate the results into a probabilistic model in order to cater for all those other scenarios that are not modelled through CFD simulations. This approach is considered an efficient means of maximizing the benefits of the CFD simulations while minimizing the need to conduct an extensive number of simulations.

5.3.4 Sensitivity to Changes in Active Safety Systems

The author (Vinnem 1997) conducted several sensitivity analyses pertaining to a detailed QRA model for leak scenarios, with the objective of evaluating the sensitivity of the QRA model to changes in the performance of active safety systems. The risk associated with personnel and material damage was analyzed, respectively. The findings of these sensitivity studies are illustrated in Fig. 5.1, indicating that the QRA model exhibits sensitivity to performance changes of the majority of active safety systems. Nonetheless, in instances where sensitivity is rather limited, the technical reasons should be examined. One notable explanation is that active safety systems are less effective in protecting against medium and large leaks, which have the highest contributions to risk.

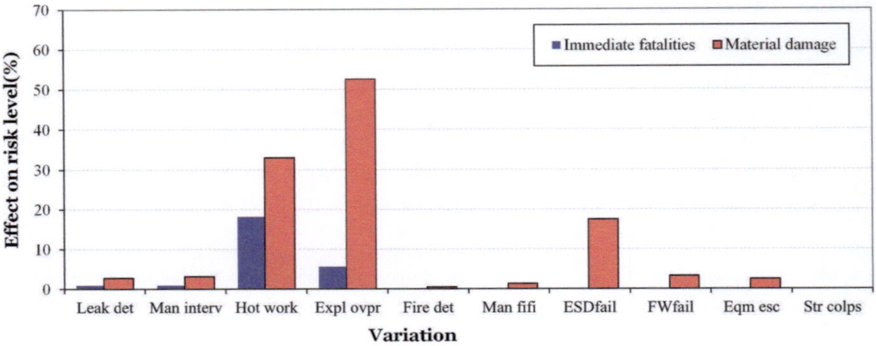

Fig. 5.1 Sensitivity analysis results of the QRA model to changes in active safety systems. *Source* Adapted from Vinnem and Røed (2020)

It can be concluded that the QRA studies should possess adequate sensitivity for the purpose of serving as a solid foundation for risk-based design.

5.4 Safety Integrity Levels for Safety Instrumented Functions

The safety instrumented systems (SISs), such as the aforementioned active safety systems, are extensively employed on offshore petroleum installations to take the process to a "safe state" upon the occurrence of potentially hazardous situations. A standard SIS typically comprises three primary components: sensors, logic solvers, and final control elements, which collectively fulfill one or more safety instrumented functions (SIFs). Sensors are responsible for measuring process variables and providing input to the logic solver, which receives signals from the sensors and decides what outputs should be sent to the final control elements. The final control elements will be activated automatically and execute the required safety actions.

The international standards IEC 61508 (IEC 2010) and IEC 61511 (IEC 2016) advocate for a risk-based method to establish the performance levels of SIFs. This is achieved by designating a safety integrity level (SIL) in the design and engineering phases of SISs. There are four different SILs, categorized from SIL1 to SIL4, which are characterized by the average probability of failure on demand (PFD). According to the IEC standards, the hazard and risk analysis should be conducted to determine:

- hazards and hazardous events associated with the process and its relevant control equipment;
- the sequence of events that may result in hazards;
- the risk in relation to identified hazards;

- the requirements for risk reduction.

To enhance the standardization and simplification of applying IEC standards in the Norwegian petroleum sector, the Offshore Norge guideline 070 (Offshore Norge 2020) has been developed. This guideline defines the minimum SIL requirements for various SIFs (e.g., emergency shutdown function, process shutdown function, etc.), which are recommended as input for QRA. This indicates that the established minimum SIL requirements can serve as a substitute for the performance requirements derived from a risk-based methodology. Table 5.5 presents the minimum requirements for some local SIFs, global SIFs, subsea SIFs and drilling BOP functions. The correlation between SIL and PFD is outlined as follows:

- SIL4: PFD ranges from 10^{-5} to less than 10^{-4},
- SIL3: PFD ranges from 10^{-4} to less than 10^{-3},
- SIL2: PFD ranges from 10^{-3} to less than 10^{-2},
- SIL1: PFD ranges from 10^{-2} to less than 10^{-1}.

5.5 Practical Considerations for Risk-Based Design

Practical considerations for risk-based design concentrate on four key areas, namely design against fire loads, explosion loads, impact from collision and impact from falling objects.

5.5.1 Design Against Fire Loads

Design against fire loads necessitates a thorough investigation and analysis of hydrocarbon loss within offshore petroleum installations, as this represents the primary source of fire risk. The hydrocarbon loss can be categorized into three main types of leaks: blowout, pipeline leaks and process leaks. These leaks will be discussed in the subsequent sections. Notably, the NORSOK Standard S-001 (Standards Norway 2018) has provided recommendations regarding fire loads that should be employed in the specification of passive fire protection for selected fire scenarios.

(1) **Blowout analysis**

Blowout represents a main source of fire risk, particularly when the wellheads are situated on or right below the surface installations. The consequences of a blowout, manifested as fire loads, are directly contingent upon the uncontrolled flow rate from the oil reservoir if well control is lost. Standard practice involves making conservative estimates when predicting flow rates. However, certain specific scenarios necessitate specialized considerations, including:

Table 5.5 Minimum SIL requirements for typical SIFs

SIF	SIL
• Process segregation through process shutdown (PSD)	1
• PSD functions: PAHH (pressure alarm high high)/LAHH (level alarm high high)/LALL (level alarm low low)	1
• PSD/emergency shutdown functions: LAHH in flare knock-out drum	3
• PSD functions: TAHH (temperature alarm high high)/TALL (temperature alarm low low)	1
• Emergency shutdown sectioning	1
• Depressurization (blowdown)	1
• Isolation of production bore upon high pressure (shut-in of one topside well)	2
• Isolation of the production/injection bore in one topside well from the production/ injection manifold/flowline	3
• Isolation of the annulus in one topside gas lift well from the gas injection manifold/line	3
• Isolation of one line of chemical injection in one topside well	2
• Isolation of riser	1
• Fire detection	2
• Gas detection	2
• Electrical isolation	2
• Release of firewater/deluge	2
• Manual initiation of fire and gas/emergency shutdown functions from the field/central control room	2
• Start of ballast system for initiation of rig re-establishment	1
• Emergency stop of the ballast system	1
• Primary and secondary barrier isolation of the production/injection bore in one subsea well from the production manifold/flowline	3
• Drilling BOP function	2

- Restricted flow versus unrestricted flow,
- Flow from tubing versus casing,
- Anticipated flow variations over time.

The design basis for safeguarding against blowout fire is primarily focused on the time needed for personnel to execute a safe evacuation, as well as the subsequent evacuation if necessary. It is impractical to design and construct extensive fire protection structures that are expected to endure a blowout fire for an extended period. Typically, the design of an external cooling system is required.

(2) **Analysis of pipeline leaks**

The pipeline leaks primarily occur in the steel pipelines close to the installation, as well as in the steel riser and flexible riser. The fire dimensions are influenced by several parameters, including:

- Pipeline diameter,
- Pipeline pressure,
- Obstructions in the flow route,
- Hole size in the pipeline.

In instances of a full-bore rupture, the fire dimensions resulting from pipeline leaks can be particularly severe. Under such circumstances, the friction loss becomes a critical factor in accurately predicting the flow rate over time.

In certain instances, accurately predicting the duration of a fire is crucial, particularly when structural fire protection is provided for a specific duration of fire exposure or when personnel may stay in a designated shelter area. The duration of the fire is influenced by several key parameters, including:

- The location of the SSIV (subsea isolation valve),
- The activation time of the SSIV,
- The hydrocarbon inventory within the pipeline.

(3) **Analysis of process leaks**

The majority of process leaks are minor leaks, which normally will not result in the accident escalation, provided that adequate safety systems are in place. Figure 5.2 shows the idealized relationship between the fire frequency and escalation probability over time. It can be observed that the fire frequency is assumed to decrease over time, whereas the escalation probability tends to increase. Thus, it is crucial to consider the safety systems carefully for a variety of leak scenarios in the process area to avoid long-duration fires. The following factors need careful consideration to ensure that the models are sensitive to reflect changes in relation to the performance of safety systems:

- Application of simplified models to complex geometries involving confinement,
- The influence of isolation and blowdown,
- The influence of deluge,
- The escalation to additional systems.

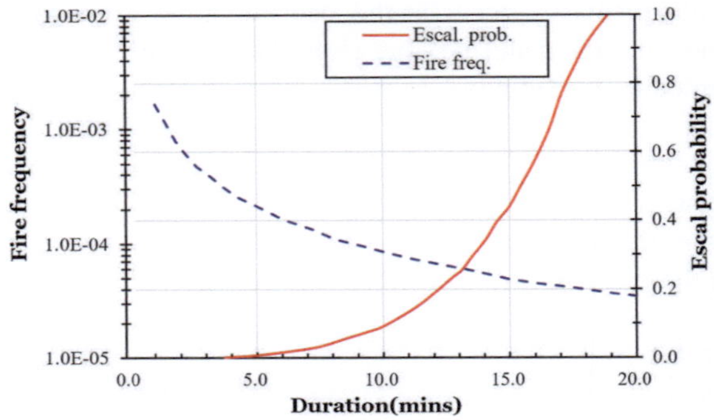

Fig. 5.2 Idealized modelling between the fire frequency and escalation probability over time. *Source* Adapted from Vinnem and Røed (2020)

With respect to escalation control within a single module, it is particularly important to provide extensive protection to prevent escalation from one segment to another. This escalation control should be based on the maximum hydrocarbon content present in any segment of the module, which will provide the basis for designing fire barriers in other areas. Furthermore, a correlation diagram illustrating the relationship between fire frequency and escalation probability over time, as exemplified in Fig. 5.2, should be developed for each leak scenario across all modules.

5.5.2 Design Against Explosion Loads

Design against explosion loads should be grounded in a probabilistic framework, as demonstrated by the extensive tests conducted as part of the BFETS program (SCI 1998). The BFETS program has proved that it is neither technically feasible nor economically viable to design in worst-case scenarios for explosions.

The frequencies for all explosion cases are typically determined through event tree analysis, and are independent of the corresponding blast loads. Explosion loads are normally calculated by professional software packages, such as Flame Accelerator Software (FLACS), which provides an extensive modelling tool for assessing the consequences of dispersion and explosions through CFD simulations within a complete three-dimensional framework for all flammable scenarios.

5.5.3 Design Against Impact from Collision

The collision accident scenarios could be generally classified into two types:

- Powered collision: a collision between the installation and passing vessels.
- Drifting collision: a collision between the installation and the nearby floating units, such as shuttle tankers, crane barges, flotels, etc.

An analysis of collision accidents over the past two decades reveals that the occurrence of collisions involving passing vessels is significantly lower than that of collisions involving attendant vessels. Consequently, some regulatory authorities have suggested that the emphasis placed on passing vessel collisions may be excessive, advocating for a greater focus on attendant vessel collisions. However, this perspective may be misleading and should not be interpreted as a justification for excluding passing vessel collisions from design considerations. A recent collision accident (AIBN 2018) demonstrates that, despite both vessels being equipped with advanced navigational systems, the risk associated with passing vessel collisions remains significant. Furthermore, it can be envisaged that the risks posed by nearby floating units can be controlled more easily than those associated with passing vessels, potentially through the implementation of operational restrictions or improved design measures.

In relation to the selection of a new site for installation, it is imperative to conduct a traffic survey to assess traffic patterns and volumes. This survey will serve as the foundation for the subsequent analysis of collision probability and consequence.

The model for collision probability could be articulated in the following manner:

$$P(\text{Collision}) = P(\text{UFM}_i) \times P(\text{Failure of Recovery}|\text{UFM}_i) \tag{5.1}$$

where

$P(\text{UFM}_i)$	Probability of the vessel's uncontrolled forward movement (UFM) type i. The UFM could be initiated in drive-off or drift-off forward conditions. For each vessel type, speed distributions should be established to distinguish between the head-on and drifting collision modes. The wind, wave, and current conditions should be considered in the drifting collision scenario.
$P(\text{Failure of Recovery}|\text{UFM}_i)$	Probability of recovery failure originating from the vessel and installation, contingent upon the vessel UFM type i.

The collision frequency model for the passing vessel collision with an installation comprises three primary elements:

- The presence of a vessel on a direct collision trajectory with the installation, which constitutes the fundamental prerequisite for a collision event.
- The capacity of the installation to implement recovery measures, primarily achieved through detection and communication methods. This could result in interception by a helicopter or standby vessel, or necessitate an emergency move-off for mobile installations.
- The capability of the vessel to execute recovery actions, which is significantly affected by the decisions made by the navigator.

Additionally, several parameters can be derived from the collision frequency model to reduce the collision risk level involving the passing vessel. Two critical parameters warrant particular attention due to their significant impact:

- Probability that the vessel is aware of the installation's presence. This can be enhanced through improved dissemination of information regarding its location and installation date.
- Probability of failure in the recovery that is initiated by the installation. This can be influenced by procedures for warning incoming vessels (e.g., a radar-based collision warning system, the utilization of visual and auditory signals on the installation, etc.) as well as by active interception measures taken by helicopters, standby vessels, or tugs.

The minimum collision design criteria for offshore structures in Norway (DNV GL 1981) are based on the impact energy. In 1981, the minimum impact energy thresholds were defined as 11 MJ for the head-on collision and 14 MJ for the sideways collision, corresponding to vessels with 5000 tons dwt. In the last 20 years it has become apparent that the size of vessels attending to the offshore installations have increased considerably in size and have higher speeds, and the approach is now to consider the actual planned use of vessels and perform a risk assessment in order to determine the expected collision energy with a frequency of 10^{-4} per year, based upon actual mass and speed distributions. Such dimensioning energies may increase to 30–50 MJ with modern vessels (Lloyd's/Safetec/DNV 2019). The determination of collision energy needs to take into account a range of collision scenarios, as well as geometrical factors and dimensions. Particularly, the frequency distribution in relation to energy can be stated for all colliding vessels collectively or for each vessel type individually.

The collision consequences typically manifest in two distinct forms of damage: global damage and local damage. The global damage refers to situations where the integrity of the installation is threatened, necessitating the evacuation of personnel and the cessation of operations. In contrast, local damage occurs without threatening the overall integrity of the installation. The collision consequences are primarily influenced by several factors, where the mass and velocity of the colliding vessel should be regarded as paramount. The relevant factors include:

- Design and dimensioning principles for the installation's structure,
- Fendering and reinforcement,
- Installation topology,
- Kinetic energy, determined by the mass and velocity of the colliding vessel,
- Distribution of collision energy between the installation and the vessel,
- Strength of the colliding vessel,
- Relative orientation of the installation in relation to the vessel,
- Specific location of the point of impact.

Apart from the primary emphasis of collision loads on the installation for the consequence analysis, several other factors necessitate careful consideration:

- The vessel's superstructure may impact the deck structure.
- The vessel may impact the risers and non-structural elements.
- The vessel may sink and impact subsea facilities.
- The vessel may catch fire, posing a threat to the installation.

5.5.4 Design Against Impact from Falling Objects

The impact from falling objects may cause serious accidents, potentially resulting in fatalities, significant structural damage, or damage to hydrocarbon systems. Notably, the fatal crane accidents in the North Sea have been infrequent since 1990. However, over the past decade, both the fatal accidents on the NCS have been caused by falling objects/loads. Hence, two primary accidental events should be addressed in the modelling of falling loads: the first involves the dropped load on the deck resulting in the damage of topside equipment and structures, while the second pertains to the dropped load entering the sea resulting in the damage of subsea equipment and structures. The former event consists solely of a single stage, specifically the descent through the air, whereas the latter event consists of three distinct stages: the initial descent through the air, the subsequent impact with the sea surface, and the final downward movement through the water.

The assessment of damage probability in association with the impact from falling objects necessitates the consideration of the following factors:

- The probability of the load being dropped from a height.
- The probability of the falling object impacting the structures and systems.
- The probability of failure pertaining to the structures and systems in response to the impact loads.

In some cases, the possibility of progressive collapse needs to be taken into account as a result of the structural damage. Furthermore, if the impact of the dropped load could result in a leak, it is imperative to conduct an analysis of hydrocarbon leak scenarios.

5.6 Questions

Q1: What are the needs and challenges of risk-based design?

Q2: Please describe the priorities for risk-reducing measures, specifically in choosing between active versus passive safety systems, and explain the rationale behind these priorities.

Q3: Please describe the priorities for risk-reducing measures, specifically in choosing between technical systems versus procedures, and explain the rationale behind these priorities.

Q4: In the context of early concept development phases, if you need to make a preliminary DiAL, would you adopt optimistic, realistic or conservative values? Please discuss and make a recommendation.

Q5: The most exact calculation of DiALs will be done with sophisticated software, which requires the availability of many design details. However, these design details are usually not known until the detailed design phase is well underway. Should you then wait for DiAL calculation until the end of the detailed design phase? Please discuss and make a recommendation.

References

AIBN (2018) Preliminary marine accident report—collision between the frigate ‘KNM Helge Ingstad’ and the oil tanker ‘sola TS’ on 8 November 2018

DNV GL (1981) Technical note—fixed offshore installations—impact loads from boats. Report no. TN–202

Havtil (2020) Regulations relating to design and outfitting of facilities, etc. in the petroleum activities. Last amended 16 December 2020

IEC (2010) IEC 61508—functional safety of electrical/electronic/programmable electronic safety–related systems. IEC, Geneva

IEC (2016) IEC 61511—functional safety—safety instrumented systems for the process industry sector. IEC, Geneva

Lloyd’s/Safetec/DNV (2019) JIP visiting vessel collision risk assessment methodology. Report no: 106074

Offshore Norge (2020) Application of IEC 61508 and IEC 61511 in the Norwegian petroleum industry (recommended SIL requirements). Guideline 070

Royal Commission on the “Ocean Ranger” Marine Disaster (1984) Report one: the loss of the semisubmersible drill rig ocean ranger and its crew. Minister of Supply and Services Canada, Ottawa, Canada

SCI (1998) Blast and fire engineering for topside systems, Phase 2. Report No. 253

Standards Norway (2010) Risk and emergency preparedness assessment. NORSOK Z-013. Rev. 3

Standards Norway (2012) Integrity of offshore structures. NORSOK N-001. Rev. 8

Standards Norway (2018) Technical safety. NORSOK S-001. Rev. 6

Standards Norway (2023) Risk and emergency preparedness assessment. NORSOK Z-013. Rev. 4

Vinnem JE (1997) On the sensitivity of offshore QRA studies. European safety and reliability conference, Lisbon, Portugal
Vinnem JE, Røed S (2020) Offshore risk assessment: principles, modelling, and applications. Fourth Edition, Springer-Verlag London Ltd

Sixth Lesson: Risk-Based Emergency Response Planning

6

What you do not prepare for becomes the crisis you cannot control.

—Herbert Hoove (1932)

Abstract

Given the severe consequences of major accidents, regulatory authorities and enterprises have paid considerable attention to emergency response planning. This lesson explores the risk-based emergency response planning from a practical point of view. To begin with, the philosophy of risk-based emergency response planning is presented. After that, particularities of offshore emergency preparedness are addressed. Then, the controversial relationship between accident prevention and emergency preparedness is explored from the perspectives of regulatory requirements and industry viewpoints. On this basis, principles for achieving a balance between regulatory and industry viewpoints are suggested. In addition, the general requirements on the establishment process of risk-based emergency response are presented on the basis of the most recent NORSOK Z-013 standard. Further, the lesson explores the area-based emergency response planning on the basis of defined situations of hazards and accidents (DSHAs), of which the critical parameters for the risk-based emergency response planning are discussed. Finally, the lesson discusses the challenges for deciding on the response level, emphasizing the determination of what is "realistic worst case" and what is not. The lesson holds critical value for readers in building a theoretical framework and practical expertise in risk-based emergency response planning.

X. Zhen and J.-E. Vinnem, *Ten Lessons on Risks*,
https://doi.org/10.1007/978-3-031-99064-9_6

6.1 Philosophy of Emergency Response Planning

Prior to delving into the main topic, the section firstly clarifies two crucial concepts pertinent to emergency response planning in accordance with the NORSOK Z-013 standard (Standards Norway 2023), as outlined below.

Emergency response	Actions taken by personnel, on or off the installation, to control or mitigate a hazardous event or initiate and execute abandonment
Emergency preparedness	Technical, operational and organizational measures, including necessary equipment that are planned to be used under the management of the emergency organization in case hazardous or accidental situations occur, in order to protect human and environmental resources and assets

The NORSOK Z-013 definition indicates that the fundamental purpose of emergency response is to mitigate the consequences when a hazardous event takes place. Furthermore, it can be noted that though the implementation of emergency response falls within the right-hand side of the bow-tie model, the emergency response planning precedes the occurrence of the hazardous event, thereby falling within the left-hand side of the bow-tie model, as shown in Fig. 1.2.

According to the risk management framework outlined in the international standard ISO 31000 (ISO 2018), as shown in Fig. 2.1, emergency response planning is an integral part of the risk treatment process.

Drawing from the aforementioned definitions and discussions, the philosophy of emergency response planning is to plan emergency actions for a diverse range of DSHAs. The Norwegian Activity Regulations (Havtil 2020) require that an emergency response be planned for:

(a) Major accidental situations;
(b) Accidental situations of limited extent;
(c) Hazardous situations implying temporary increase in risk.

There exists a crucial premise underlying the emergency response planning, i.e., its necessity cannot be decided upon the results of a risk assessment. In other words, emergency response planning cannot be neglected due to the low probability of hazardous and accidental situations occurring. This further underscores the obligation to provide emergency response plans even though the risk level of the hazardous and accidental situations is extremely low.

Apart from this crucial premise, the philosophy of emergency response planning focuses on establishing its critical parameters in a risk-based context. These parameters include the response time and the required capacity for emergency response actions. This indicates that the parameter values are determined on the basis of the worst-case situations that could reasonably occur, as well as modelling of accident scenarios in a risk assessment approach.

Notably, such a distinction on the philosophy of emergency response planning is agreed by experts from Havtil: the necessity of emergency response planning cannot solely be based on risk assessment, whereas its critical parameters can be determined in a risk-based context.

In the final, the NORSOK Z-013 standard (Standards Norway 2023) outlines the principles for emergency preparedness, which should be regarded as minimum requirements:

- Establishment of the DSHAs shall be based on risk assessments.
- Emergency response is normally not dimensioned to handle two independent DSHAs at the same time. However, it should be considered if one DSHA may trigger another DSHA.
- Emergency response shall adhere to a prioritization principle, with respect to people, environment, asset and reputation (PEAR) in that order;
- Personnel should not be exposed to unnecessary danger. When using emergency response personnel for rescue or combat, the potential risks of injury or life loss shall be assessed.
- The emergency response organization shall be dimensioned, adhering to the robustness principles.

6.2 Particularities of Offshore Emergency Preparedness

The particularities of offshore emergency preparedness should be addressed in the risk-based emergency response planning, as the development of offshore petroleum fields represents a complex and multi-dimensional systems engineering challenge. Notably, the remoteness and need for self-sufficiency are the most particular characteristics concerning offshore emergency response planning.

- Offshore remoteness

  Offshore installations located in deepwater are typically situated more than 100 km from the coastline. This remoteness indicates that in the case of an emergency, the installation must be self-sufficient in relation to resources and on-scene emergency management. Additionally, the remoteness indicates that helicopters cannot be depended upon for immediate assistance in an emergency, as onshore helicopters typically require a minimum of 60 min to respond after scrambling. Furthermore, certain accident scenarios may restrict the use of helicopters.
- Harsh environmental conditions

  The harsh environmental conditions may be characterized by strong winds, huge waves, great water depths, low temperature, sea ice, and other related factors. The harsh environmental conditions may restrict the use of some critical emergency response resources.
- Offshore installation types

Offshore installations come in many types. Smaller ones are often unmanned and can be interconnected by bridges. Large installations may house hundreds of personnel. Some installations are fixed to the seabed, such as jackets and jack-ups, while others float on the sea surface, such as semi-submersibles and FPSOs. Different types of offshore installations may have varying requirements for emergency preparedness measures.

6.3 Accident Prevention Versus Emergency Preparedness

From a philosophical point of view, the relationship between accident prevention and emergency preparedness is an interesting topic and merits in-depth examination. Notably, the petroleum enterprises have for some time expressed the view that improved accident prevention is not given any credit by the regulatory authorities for emergency preparedness. This section aims to explore this interesting topic from the perspectives of regulatory requirements and industry viewpoints, and then suggests principles for achieving a balance between accident prevention and emergency preparedness, based on an extensive evaluation of offshore emergency preparedness by the author (Vinnem 2011).

6.3.1 Regulatory Requirements

The Norwegian and UK regulations governing offshore facilities and operations include important requirements that influence the balance between accident prevention and emergency preparedness. However, these regulations do not provide explicit requirements that could be used to determine what this balance should be.

Norwegian regulations have requirements that call for continuous risk reduction, irrespective of the risk level. There should be efforts made to reduce risk through preventative or mitigation measures, and good proposals shall be implemented unless the negative effects of such proposals, including costs, are significantly disproportionate to the benefit. This principle is known as ALARP. The UK has established similar requirements for ALARP evaluations and the risk reduction process.

In addition, both Norwegian and UK regulations have some minimum requirements for accident prevention and emergency preparedness. For example, Norwegian regulations have requirements regarding the use of free-fall lifeboats, along with two independent systems for rescuing personnel who fall into the sea. There is also the requirement that when the dimensioning of passive fire protection is performed, this shall be done without consideration of the effect of active fire protection, such as fire water.

There are further requirements in the Norwegian regulations that personnel may be evacuated and rescued in major accidental situations defined in QRA studies, accidental situations of limited extent and situations implying a temporary increase of risk.

In the UK , there is a requirement to provide arrangements that give a "good prospect of escape, evacuation and recovery" for all foreseeable events (HSE 1995).

Generally, there is a tendency in regulations to call for adequate emergency provisions for foreseeable situations, regardless of what has been done so as to reduce the probability of these situations occurring. Consequently, there is no right to reduce the extent of emergency preparedness provisions due to the improved preventative safety.

It has been noted that enterprises that have presented arguments to Norwegian and UK authorities regarding the reduction of emergency preparedness due to improved preventative safety have received responses indicating that such an approach is not considered viable.

6.3.2 Industry Viewpoints

Several international and national standards for the offshore industry (NORSOK 2006; ISO 1999, 2000) highlight the fundamental principle of risk reduction, stating that priority should be given to probability-reducing measures rather than consequence-reducing measures for accidents. These standards are all referenced in Norwegian legislation. The implication of such a principle is that if the occurrence probability of accidents could be reduced, there would be no need to enhance emergency preparedness. This viewpoint is commonly held within the industry, based upon the wording of the standards.

Nonetheless, there could be other priorities and circumstances that will inspire the industry to enhance the emergency preparedness arrangements. One important circumstance is the need to maintain a good reputation. This is a significant factor, particularly in the context of emergency preparedness in relation to environmental spills. For instance, Norway has been engaged in a fundamental discussion regarding the possible permission to start exploration and production activities in certain environmentally sensitive regions in northern Norway. These regions experience conditions that are not fully arctic, but are close to arctic conditions. The debate is focused on environmental spills and associated risks, and the willingness to invest in the improvement of preventative measures and emergency preparedness is present. This willingness is driven not only by regulatory requirements and pressure from non-governmental organizations, but also by the need to maintain a good reputation.

6.3.3 Proposed Principles for Balance Between Accident Prevention and Emergency Preparedness

The balance between accident prevention and emergency preparedness is hard to achieve in a straightforward way. A critical question arises: Is there a practical way out of the dilemma, one that may provide a compromise between the regulatory and industry viewpoints? The authors try to outline the principles so as to reach such a compromise through a four-step process. Notably, these principles are only applicable within the framework of the risk-based legislation, rather than the prescriptive legislation.

Step 1: The regulatory frame conditions must be satisfied at all times. This includes:
- Minimum requirements for preventive safety,
- Minimum requirements for emergency preparedness,
- Minimum requirements regarding acceptable risk levels, i.e., upper tolerability limits.

Step 2: The ALARP process ought to be implemented according to regulations.

Step 3: When making decisions regarding measures for risk reduction:
- Priority should be given to probability-reducing measures over consequence-reducing measures.

Step 4: In cases where specific emergency preparedness measures are not implemented, the preferred preventative measures must be specified and implemented.

6.4 How to Establish Risk-Based Emergency Response?

While the QRA details the potential hazards and consequences that could arise on an offshore installation, the emergency response planning focuses on how these challenges could be dealt with in practice.

Figure 6.1 presents the emergency preparedness assessment process in the NORSOK Z-013 standard (Standards Norway 2023). It can be seen that the establishment of emergency response involves five critical stages:

- Input from risk assessment,
- Establishing the context,
- Emergency preparedness analysis,
- Emergency response plan,
- Emergency training and exercise.

In the subsequent sections, each stage with respect to the establishment of emergency response will be presented and analyzed from a practical perspective. Particular attention will be given to aligning the emergency preparedness analysis process with the most recent edition of the NORSOK Z-013 standard (Standards Norway 2023), reflecting current best practices.

Notably, it can be inferred from Fig. 6.1 that the primary purpose of the emergency preparedness assessment (Standards Norway 2023) is to provide input to the establishment or update of the emergency response plan by:

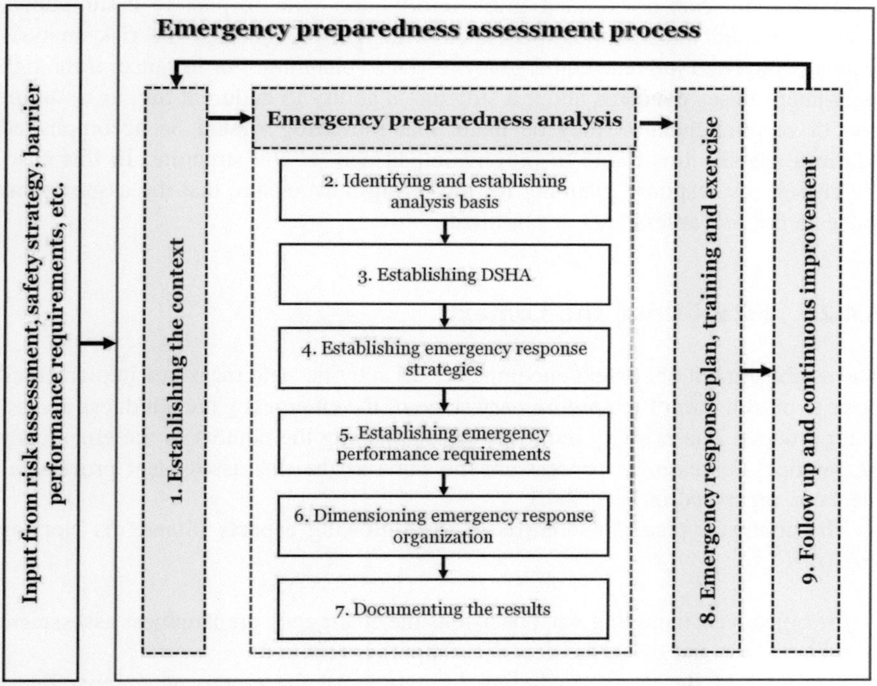

Fig. 6.1 The process for emergency preparedness assessment. *Source* Adapted from NORSOK Z-013 (2023)

- establishing emergency response strategies;
- dimensioning a robust emergency response organization;
- providing input to plans for training and exercises.

6.4.1 Input from Risk Assessment

A primary objective of risk assessment is to provide input to the emergency response planning, as depicted in Fig. 6.1. The risk assessment identifies and describes potential hazards, accident scenarios and consequences specific to the installation, providing pertinent information for emergency response planning. The description of accident scenarios includes the comprehension of the barrier functions, as well as the barrier elements in place and the corresponding barrier performance requirements to achieve the barrier functions. It is also important to comprehend accident scenarios in which barrier elements failed to function as intended, such as when isolation is not accomplished or when gas detection does not work.

In addition, risk assessment gives information with respect to assumptions, boundaries, and uncertainties associated with risk models and the risk analysis results, which can influence emergency response planning. For instance, if the risk assessment raises concerns about a structure's ability to endure a fire, an assumption or recommendation may be made that mustering should be accomplished within a specific timeframe to prevent impairment of the structure. In this case, the emergency response planning for mustering must ensure that the assumptions made in the risk assessment are fulfilled.

6.4.2 Establishing the Context

The establishment of context encompasses all activities and measures implemented prior to or as a part of the preliminary stage of the emergency preparedness assessment process. The primary purposes of establishing the context in the emergency preparedness assessment process are the same as the risk assessment process, as has been presented in Sect. 2.4.1.

The context typically encompasses the following aspects (Standards Norway 2023):

- the objectives, including what decisions the emergency preparedness assessment will support and when the decision support is required;
- definition of the scope, including limitations of the emergency preparedness assessment;
- description of the facility;
- a set of principles for emergency preparedness;
- description of the regulatory requirements;
- relevant standards and guidelines.

In practice, the emergency preparedness analysis is carried out through a series of workshops conducted by a workgroup. It is crucial for this workgroup to receive support from critical stakeholders. The members of the workgroup must possess the competence to comprehend how the hazards identified in the risk assessment may evolve. The workgroup ought to possess the authority to determine an adequate level of emergency preparedness. Generally, a workgroup comprises the offshore installation manager, a safety representative, a process operator, a drilling/well expert, and a nurse. It is important for the workgroup to have a clear understanding of the organization, including roles and responsibilities, as well as the technical safety systems that affect the development of accident scenarios.

Here are some examples of background information to consider:

- relevant regulatory requirements, guidelines and standards;
- relevant information from the risk assessment,
- existing emergency preparedness analysis, if available,
- design drawings and a 3D design model,

- agreements for collaboration with emergency response resources,
- company-specific governing documents,
- experience from emergency response training and exercises,
- information regarding the technical conditions of the facility, including the status of barrier elements.

6.4.3 Emergency Preparedness Analysis

As outlined in Fig. 6.1, emergency preparedness analysis consists of six key steps, and each step is introduced as follows, based on the NORSOK Z-013 standard (Standards Norway 2023).

Step 1: Identifying and establishing an analysis basis

In comparison with the previous edition of NORSOK Z-013 standard (Standards Norway 2010), this step is separated from the step of establishing the context and functions as a standalone procedure. The basis of the emergency preparedness consists of a description of the context, a description of assumptions, identification and assessment of learning points, as well as personnel and competence requirements.

The assumptions should be clearly and explicitly documented. Examples of learning points include relevant experience, lessons from similar events, technological development, and new knowledge.

Step 2: Establishing DSHAs

The selection of DSHAs encompasses the following situations of hazards and accidents:

- major accidents, including the dimensioning of accidental events identified in the risk assessment;
- accidental events identified in the risk assessment that are not classified as major accidents, but still pose separate challenges to emergency preparedness;
- temporary increase of risk or incidents not covered in the risk assessment, such as the hazards of drifting objects and man overboard, etc.;
- events that have been identified in comparable activities.

The rationale behind this selection of DSHAs is that, beyond major accident hazards explicitly analyzed in the risk assessment, there may be other hazards that have not been thoroughly analyzed in the risk assessment but should still be considered in the emergency preparedness analysis. Examples of these hazards include fire in the living quarters and spills of chemical compounds.

Overall, it is expected that all potential hazards that could pose a challenge to the emergency response organization are incorporated into the emergency response planning. This encompasses hazards characterized by a low-risk contribution due to low probabilities, such as less than 10^{-4}, as well as hazards that have significant uncertainties.

Step 3: Establishing emergency response strategies

This step necessitates the development of two primary strategies: (1) mustering and evacuation strategies, (2) emergency response strategies specific to each DSHA.

For each accident scenario, it is crucial to discuss and determine the emergency response strategies, as well as allocate the resources. Resources can encompass operational, organizational and technical aspects. To ensure all these resources are taken into account, a useful question of "who does what with which equipment?" has been developed. In this question, "who" refers to the organizational barrier element, "what" pertains to the operational barrier element, and "with which equipment" addresses the technical barrier element.

Further, it is essential to consider potential split-scenarios, where personnel may be unable to access the primary evacuation means. In such instances, alternative evacuation means or operational restrictions are necessary.

Step 4: Establishing emergency performance requirements

Emergency performance requirements should be established for the facility and the DSHAs, on the basis of regulations, authority guidelines, classification society rules, standards, the enterprise's emergency preparedness principles and requirements.

Emergency performance requirements can be exemplified by the necessity to rescue an individual who has fallen into the sea within a designated timeframe. Emergency performance requirements can be established according to capacity and functionality (for instance, two individuals having the necessary competence), availability (such as 24/7 throughout the year), and response time (within a specified number of minutes).

Step 5: Dimensioning the emergency response organization

The objective of dimensioning an emergency response organization is to identify the minimum required functions for a robust emergency response organization and the equipment required to fulfill the functions.

It is the responsibility of the operating company to dimension the emergency response to ensure that representative situations are handled in accordance with the performance requirements established by the work group. As part of this work, it is crucial to evaluate whether there are any incompatible functions, such as the personnel being assigned to multiple tasks that need to be performed simultaneously.

In hindsight, it can be frequently observed that when an emergency organization is not adequately dimensioned for a situation, the inadequacies can be attributed to two primary reasons. On the one hand, the actual scenario was far more severe than the representative scenario set up in the emergency response plan. On the other hand, the performance requirements established were not sufficiently robust to handle the event that occurred. The root causes can be exemplified by a lack of knowledge among the workgroup regarding the potential scenarios and their possible development, as well as the absence of authority among workshop members to approve the established performance requirements. At times, the management did not view the emergency response analysis as the only foundation for emergency response planning, which led to the decision not to approve the implementation of the recommendations provided by the workgroup.

Step 6: Documenting the results

According to the NORSOK Z-013 standard (Standards Norway 2023), documenting the results should primarily cover the context description, a description of the methodology employed and the work process (including participants), descriptions of all DSHAs, emergency response strategies, performance requirements for emergency response, the recommended dimensioning of the emergency response organization, a description of differences in viewpoints and interpretations within the workgroup, and recommendations.

6.4.4 Emergency Response Plan

The emergency response plan delineates the practical execution of the emergency response. The document of the emergency response plan is designed to serve as a tool for the personnel involved in the emergency response, and thus should be concise and to the point. The involved personnel ought to possess the necessary competence and training, be well acquainted with their own duties as well as the roles and responsibilities of other involved parties. This indicates that explanations for the implementations of measures should be excluded to avoid resulting in an overly detailed document. In many instances, the emergency response plan can incorporate specific checklists tailored to each function for emergency response. These checklists ought to encompass all phases of emergency response, including detection and alert, danger limitation, rescue, escape and evacuation, normalization.

In the emergency response plan, the tactical, operational and strategic levels are typically differentiated, and are often referred to as the first line, second line and third line, respectively. The first line pertains to the emergency response organization on the offshore installation. The second line involves personnel in an onshore emergency response room, who provide support to the offshore organization and external resources, including helicopters, vessels, etc. The third line is concerned with the company management, who are responsible for managing media communications, engaging with external companies, etc.

6.4.5 Emergency Training and Exercise

Training and exercise are essential for equipping emergency response personnel with the necessary knowledge and skills to effectively perform their duties in an emergency response situation.

During emergency response training, unskilled personnel often engage themselves in software systems and other structural elements, which can hinder their ability to obtain a comprehensive perspective and anticipate potential challenges. To develop a sufficient understanding of the situation and prepare for unknowns, it is essential for the knowledge to be internalized. Training and exercises are crucial for achieving this goal.

For emergency coordinators and other off-site members of an emergency response team, effective training can be conducted without the need for actual resources, such as helicopters. By building a virtual and simulation training platform, the emergency response team can communicate with simulated agents that represent real resources. This training method has been proven effective in practice while minimizing resource expenditure during the exercise.

When selecting scenarios for training, it is important to choose from those determined in the emergency preparedness analysis. This selection ensures that the scenarios being exercised are representative. When emergency response personnel receive training focused on representative events, it is probable that they will also be able to employ the knowledge and skills acquired to address other accidental events. It should be noted that the actual situation will probably not be the same as the selected training scenarios. However, when the personnel within the emergency response organization are well aware of their specific roles and responsibilities, and have received training to cope with representative scenarios, they will probably be able to adapt as needed by the actual situation.

6.5 Area-Based Emergency Response Planning

The Norwegian framework regulations (Havtil 2019) require operators to coordinate and cooperate on the offshore emergency response. Hence, the area-based emergency response (Offshore Norway 2015) is established with coordination and cooperation between operators regarding the available resources, including search and rescue (SAR) helicopters, emergency response and rescue vessels (ERRV) and marine traffic surveillance. So far, Norway remains the only country that has formalized emergency preparedness cooperation and coordination into alliances within clearly defined geographical areas. Within each alliance, there are shared emergency response resources. To date, a total of six formal alliances have been

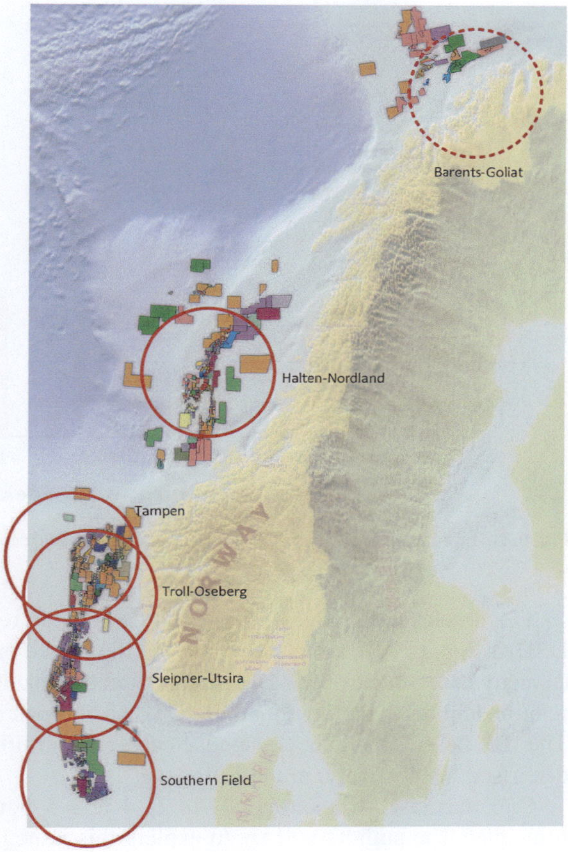

Fig. 6.2 Overview of the area-based emergency response alliances on the NCS, indicating with solid circles

established (Johansen et al. 2020), as indicated in Fig. 6.2. There are five solid circles and one dotted circle, which recently has been upgraded to a full alliance.

The following DSHAs serve as the foundation for area-based emergency response planning. Their typical frequencies on the NCS are presented in Fig. 6.3. Notably, these frequencies are roughly estimated.

DSHA 1: Personnel overboard when working over the sea.
DSHA 2: Personnel requiring rescue following a helicopter ditching.
DSHA 3: Persons requiring rescue following emergency evacuation to sea.
DSHA 4: Collision hazard.
DSHA 5: Acute oil spill.
DSHA 6: Installation fire requiring external firefighting.
DSHA 7: Acute injuries or illnesses requiring external medical response.
DSHA 8: Helicopter crash on the installation, resulting in severe injuries.

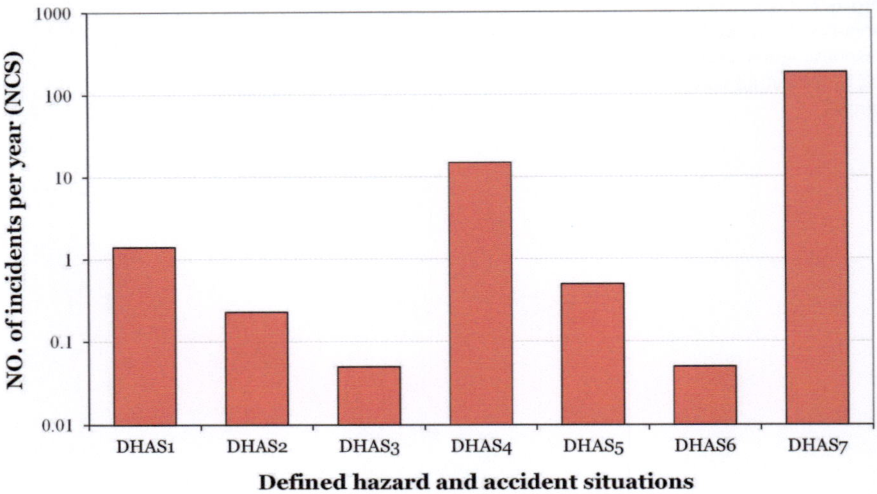

Fig. 6.3 Typical frequencies for DSHAs on the NCS. *Source* Adapted from Vinnem (2012)

The crucial parts of the area-based emergency response planning, specifically DSHAs 2, 3, 6, 7, 8, are grounded in the risk-based principle. Thereinto, the rescue capacity assessment of the DSHA 3 is directly linked to risk analysis, whereas the relationship between the remaining DSHAs and risk analysis is somewhat indirect. The risk-based emergency response capacity and time for the relevant DSHA are discussed, respectively, in the following.

One may ask, why are DSHAs 1, 4 and 5 not grounded in the risk-based principle? The authors will try to explain this issue.

The DSHA1 is typically referred to as "Man overboard", but is limited to scenarios in which personnel are working over the sea. When personnel are working over the sea, they are required to have surveillance such that if a person falls overboard, it shall be detected immediately, and the person shall be rescued within maximum 8 min in order to avoid hypothermia. This has been formulated as a prescriptive requirement for work offshore, thus not dependent on any risk analysis.

The consequences of a merchant ship colliding with an installation can be catastrophic. So even if the probability is low, it has been considered that traffic monitoring needs to be available for all installations, irrespective of the probability, in order to prevent the collision hazard (DSHA4) from occurring. Therefore, it is not dependent on risk levels and is not risk-based. This is implemented with two traffic surveillance centers that together cover the entire Norwegian sector.

The consequences of a large acute oil spill (DSHA5, from a blowout or other sources) can be catastrophic for the marine environment, especially in the cold waters of the Norwegian sector. So even if the probability is low, it has been considered that emergency resources to combat oil spills need to be available for all installations, irrespective of the probability. Therefore, the emergency preparedness is not dependent on risk levels and is not risk-based. This DSHA is addressed by providing standby vessel in the areas with oil spill combat equipment.

6.5.1 Risk-Based Rescue Capacity After Helicopter Ditching

Risk analysis of helicopter transportation of offshore employees from the shore base to offshore installations shows that the most likely scenario where survivors need to be rescued is a helicopter ditching either en route to the installation or during landing on the installation, with up to a full helicopter with 19 passengers and 2 pilots. In this case, the maximum number of individuals onboard is 21, including 2 pilots. Consequently, the rescue of 21 persons within a certain time limit is the requirement.

The required time limit, specifically the response time, is contingent upon the capabilities of the survival suits worn by passengers. The adopted approach incorporates a safety factor of 1.5, suggesting that if the designated response time is 120 min, the capabilities of the survival suits must be validated for a minimum duration of 180 min. This requirement does not pose a problem in the North Sea and Norwegian Sea, as the water temperature does not fall below 2 °C. In accordance with Offshore Norway requirements, the survival suits employed in Norway must be tested to provide protection against hypothermia for a duration of 6 h.

The requirement is thus to rescue a maximum of 21 persons within a time frame of 120 min.

6.5.2 Risk-Based Rescue Capacity After Emergency Evacuation

The risk-based rescue capacity after emergency evacuation is required. What needs to be emphasized is that the requirements for providing risk-based rescue capacity are not directly based on risk assessment results, but rather on the principles of risk-based dimensioning. This is due to the differences among installations concerning their safe evacuation provisions.

To ascertain a risk-based rescue capacity, the following steps are carried out:

Step 1: A review of the assumptions, premises, and results derived from the QRA is conducted to ascertain the requirements necessary for calculating risk levels. These assumptions and premises may embody critical conditions necessary for the risk level to be deemed tolerable.

Step 2: The maximum number of personnel, as derived from the accident scenarios outlined in Step 1, serves as the minimum requirement for rescue capacity.

Step 3: The potential number of personnel (based on the QRA), who may need rescue, is determined based on the escape and evacuation arrangements available on the installation.

If the number of personnel in Step 3 is higher than the minimum capacity from Step 2, the number of personnel from Step 3 is the dimensioning. Otherwise, the number from Step 2 is retained.

6.5.3 Risk-Based External Fire Fighting

The capacity of external firefighting must be dimensioned in accordance with assumptions established during the risk analysis. These requirements are usually defined for standby vessels within a certain response time.

Requirements for response time must be determined based on assumptions and requirements in a QRA.

6.5.4 Risk-Based External Medical Response

It is possible to perform a risk assessment focusing on occupational accidents on installations and consider statistics about persons who become seriously ill. This has not been done, but more general statistics have been considered for the entire Norwegian offshore sector. These statistics show that three or four serious occupational accidents occur each month. Each month, there are also several cases of critical illness that necessitate rapid medical assistance in a hospital to ensure survival.

The medical requirements for seriously ill and injured personnel involve ensuring the availability of advanced thrombolytic equipment within an hour and transporting such persons to an onshore hospital within a 3-h timeframe (Offshore Norway 2015).

6.5.5 Risk-Based Rescue Capacity in Helicopter Accidents

There are two requirements associated with helicopter accidents, the first one is as discussed above, the rescue of survivors in the sea following a helicopter accident on water, and the second is the medical capacity to treat survivors following a helicopter accident on the installation.

A risk analysis of helicopter accidents on the installation has been performed (Vinnem 2012). This risk analysis has determined that the following is the dimensioning scenario:

- Helicopter toppling over on the helideck, followed by a fuel fire

This scenario is analyzed to result in the following consequences for personnel in the helicopter:

- One fatality, three seriously injured persons and four persons with minor injuries

The required rescue capacity for a helicopter crash on the installation is, therefore, to attend to three seriously injured personnel and four persons with minor injuries. There is no time requirement associated with this scenario.

6.6 Challenges for Deciding on Response Level

Section 6.1 argues that low probability cannot be the argument to disregard a scenario for emergency planning. Yet no one has required that emergency response planning should be dimensioned for the scenario with a meteor crashing on an offshore installation. A meteor landing on an installation is not an impossible scenario; meteor landings on land and on sea occurs globally a few times each year, if they do not burn up completely when entering the atmosphere. In theory, it could happen on an offshore installation.

Another scenario in the North Sea is a passenger aircraft crashing on an offshore installation. There are several aircraft crashes into the ground each year, and it could in theory happen on an offshore installation. Some people claim to have roughly "guesstimated" the probability of such a scenario to be on the order of one in a million years, 10^{-6} per installation per year. Yet no one has emergency response planning for such a scenario.

This actually means that the so-called "worst case" is not really the "ultimate worst case", but some kind of "realistic worst case". But what is "realistic"? If it has never occurred anywhere globally in the offshore industry, is it still realistic?

In 1988, when the Piper Alpha disaster occurred in the North Sea region of the UK, a similar accident had never occurred anywhere before. The emergency response planning had not covered such a scenario, and a death toll of 167 persons resulted. The same with the Alexander Kielland accident in the Norwegian North Sea sector in 1980, a similar accident had not occurred anywhere before. Despite adequate emergency response planning being in place, the accident resulted in a high death toll of 123 persons. The emergency response planning was criticized in both cases.

This is why the "realistic worst case" has to include scenarios that no one has experienced but are covered in risk analysis under the most severe circumstances. However, even the risk analysis would not include a crash of a meteor or a passenger aircraft into the installation. So, in practice, there exists a certain limit somewhere, but it is probably an unknown limit.

The challenge for emergency response planning is to determine what is the "realistic worst case" and what is not. It is important to be aware of this challenge, but there is unfortunately no exact answer to this challenge.

6.7 Questions

Q1: Which critical situations are prioritized for emergency response planning?

Q2: What are the advantages of the area-based emergency response cooperation?

Q3: What are the limitations for each area (see Fig. 6.2) with area-based emergency response cooperation?

Q4: In which scenarios could there be an upper limit on the number of installations could belong to an area-based cooperation?

Q5: Why should not two helicopters landing controlled on water be the required capacity of area-based emergency assistance for helicopter accidents?

References

Havtil (2019) Regulations relating to health, safety and the environment in the petroleum activities and at certain onshore facilities. Last amended 26 Apr 2019

Havtil (2020) Regulations relating to conducting petroleum activities. Last amended 16 Dec 2020

HSE (1995) Prevention of fire and explosion, and emergency response regulations. HMSO, London

ISO (1999) Control and mitigation of fires and explosions on offshore production installations-requirements and guidelines. ISO 13702. International Standards Organization, Geneva

ISO (2000) Guidelines on tools and techniques for identification and assessment of hazards. ISO 17776. International Standards Organization, Geneva

ISO (2018) Risk management—guidelines. ISO 31000. International Standards Organization, Geneva

Johansen TS, Ranum SA, Wold MS, Vinnem JE, Lindberg M, Jacobsen SR, Steinbakk AH (2020) Emergency response alliances on the Norwegian continental shelf. In: ESREL 2020-PSAM, vol 15, pp 4318–4325

Offshore Norway (2015) Recommended guidelines for establishment of area-based emergency preparedness

Standards Norway (2006) Technical safety. NORSOK S-001

Standards Norway (2010) Risk and emergency preparedness assessment. NORSOK Z-013. Rev.3

Standards Norway (2023) Risk and emergency preparedness assessment. NORSOK Z-013. Rev.4

Vinnem JE (2011) Evaluation of offshore emergency preparedness in view of rare accidents. Saf Sci 49:178–191

Vinnem JE (2012) Area based emergency response planning on the Norwegian continental shelf, supporting documentation of premises and scientific assessments in offshore Norway, Guideline No. 64. Report No.2011100-02

7 Seventh Lesson: Barriers for Major Accident Risk

If you think safety is expensive, try an accident.
—Stelios Haji-Ioannou (2004)

Abstract

The term "barrier" has emerged as a prevailing and prominent concept in the field of offshore risk management. It is frequently employed in various contexts, with significantly different meanings, such as human barrier, operational barrier, organizational barrier, technical barrier and non-technical barrier. This variability suggests that achieving a consensus on a clear and uniform interpretation of the barrier concept is challenging. This lesson aims to clarify the concepts of the barrier and barrier management, along with the principles and process involved in barrier management. The foundation for this clarification is the set of guidance documents titled "Principles for Barrier Management in the Petroleum Industry" (Havtil in Principles for barrier management in the petroleum industry. Stavanger, Norway, 2013; Havtil in Barrier memorandum: principles for barrier management in the petroleum industry. Stavanger, Norway, 2017). In addition, this lesson addresses the regulatory requirements pertaining to barriers, as well as the structure principle of barriers and hazards. Lastly, the challenges faced in implementing barrier management are discussed.

7.1 Barrier and Barrier Management Concepts

This section aims to clarify the barrier and barrier management concepts. Particularly, the rationale behind barrier management is also discussed.

X. Zhen and J.-E. Vinnem, *Ten Lessons on Risks*,
https://doi.org/10.1007/978-3-031-99064-9_7

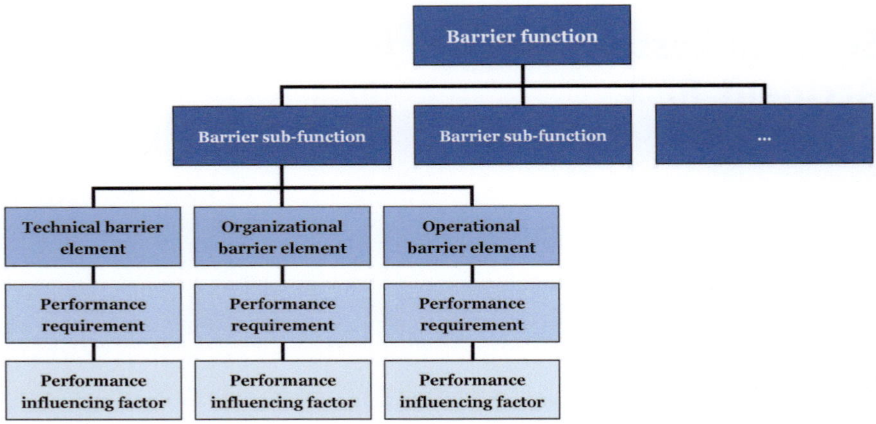

Fig. 7.1 A generic example of a barrier hierarchy. Reprinted from Barrier memorandum 2017, with permission from Havtil

7.1.1 What Are Barriers?

Barrier is defined as a measure intended to identify conditions that may lead to failure, hazard and accident situations, prevent an actual sequence of events from occurring or developing, influence a sequence of events in a deliberate way, or limit damage and loss. The barrier concept involves multiple hierarchical levels, as illustrated in Fig. 7.1. The corresponding definitions within the barrier hierarchy are interpreted in the following.

Barrier function	The barrier function refers to the task or role of a barrier (i.e., what the barrier should do). The realization of a barrier function involves a hierarchy consisting of: • The function at the highest level and sub-functions below this, if necessary • Elements (specific equipment, personnel and operations designed to take care of the function) • Performance requirements (verifiable requirements for the properties of the barrier elements in order to ensure that the barrier is effective) • Performance-influencing factors (factors identified as having significance for barrier functions and the ability of barrier elements to function as intended)
Barrier element	Technical, operational and organizational measures or solutions involved in the realization of a barrier function. For instance, technical barrier elements might be sensors that measure the pressure in a well, while organizational and operational barrier elements might be mud loggers and drillers who monitor, detect and implement measures in the event of loss of well control

(continued)

(continued)

Technical barrier element	Equipment and systems involved in the realization of a barrier function
Operational barrier element	The actions or activities that personnel must perform in order to realize a barrier function
Organizational barrier element	Personnel with defined roles or functions and specific competence involved in the realization of a barrier function

It is important to note that the term "barrier system" is also employed in some cases to describe a system that has been designed and implemented to perform one or more barrier functions.

Drawing from the aforementioned definitions, the implications of barriers, performance requirements and performance-influencing factors are further elaborated below.

Barriers

In the Barrier memorandum (Havtil 2017), the distinction between "safe and robust solutions" and "barriers" has been made. The Management Regulations distinguish between two concepts: (1) establishing a solution that is optimal for avoiding getting into failure, hazard and accident situations (Sect. 7.4 of the Management Regulations concerning risk reduction) and (2) establishing barriers intended to prevent failure and hazard situations from developing into incidents, as well as limit loss, harm and disruption (Sect. 7.5 of the Management Regulations concerning barriers). Thus, barriers supplement a safe and robust solution, as exemplified in Fig. 7.2. The motivation of this distinction is to clarify what barriers are and what they are not. Barriers must detect incipient incidents, prevent the development of a chain of events and limit harm and disruption. The elements of a barrier in combination must be able to realize this function.

The Management Regulations require that performance requirements are defined for barrier elements, and further, it must be possible to verify this performance in order to be aware of which barrier elements are not functioning and are impaired. There is a widespread perception that anything that may help prevent an incident from occurring or limit damage is a barrier. However, this overly broad interpretation makes it difficult to define verifiable performance requirements.

Barrier elements should be restricted to those measures or solutions that have a specific role in ensuring the barrier function. For example, a procedure may be a useful aid to describe what is needed to safeguard a barrier function. However, the procedure itself cannot be considered a barrier element.

Performance Requirements

To ensure that the barrier function works as intended in the specific risk picture, it is important to define specific and verifiable requirements for the performance of barrier elements, as well as the barrier function, where appropriate. In this context, "performance" refers to the properties that a barrier element must possess in order to ensure that the barrier is effective.

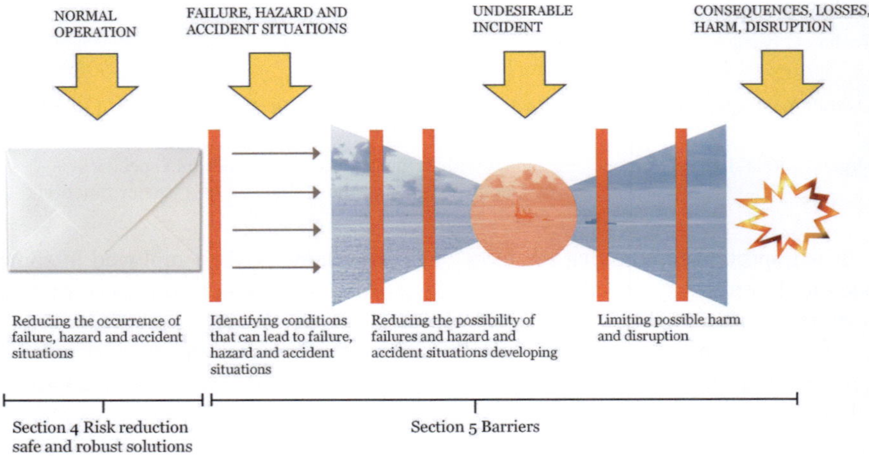

Fig. 7.2 The barrier diagram shown functions (in red) to handle failure, hazard and accident situations outside of normal operations. Adapted from Ersdal (2014). Reprinted from Barrier memorandum 2017, with permission from Havtil

Performance requirements may include requirements for functionality, integrity and robustness as follows.

- Functionality: the effect that the barriers have on the sequence of events.
- Integrity: the ability and potential of the barriers to be in place and intact at all times.
- Robustness: the ability of the barriers to withstand situations that are somewhat different from the intended circumstances and to "survive" a failure, hazard and accident situation.

It is important to recognize that the terms "functionality", "integrity" and "robustness" are used here as general concepts. According to the international standard ISO 13702 (ISO 2015) and Norwegian management regulations, "Functionality" refers to the performance of barriers in terms of their capacity and effectiveness. Furthermore, the aspects of barrier performance pertaining to reliability, availability and integrity are grouped together under the category of "integrity". Additionally, the barrier performance related to the ability to resist loads and maintain robustness is classified under "robustness".

Table 7.1 presents examples of performance requirements for the barrier function "safe evacuation" and its associated barrier elements.

In general, performance requirements are dimensioned relative to the most serious incident. This implies that there is a requirement for the largest capacity and shortest mobilization time.

Further, performance requirements will depend on site-specific factors, the incidents and their seriousness. Performance requirements may not necessarily be limited

Table 7.1 Examples of performance requirements for the barrier "safe evacuation"

Barrier	Performance requirements
Barrier function	Time to safely evacuate the personnel
Technical barrier element	Capacity of means of evacuation, audibility of alarms, etc.
Organizational barrier element	Specific competence requirements for personnel, such as frequency of participation in safety drills, etc.
Operational barrier element	Criteria for warning or informing, such as time requirement, etc.

to numerical values, but may also be qualitative assessments. This is perhaps especially relevant when defining performance requirements for personnel and the actions they are to perform.

Performance-Influencing Factors

Many factors will affect the performance of a barrier. However, not all factors are equally important. Thus, it may be important and appropriate that performance-influencing factors are restricted to those that have a significant effect on the barrier elements' ability to function.

In order to identify key performance-influencing factors, it's useful to review what may affect specific barrier elements and functions. Interviews with maintenance personnel and operators may be useful for obtaining important information. Reviews of incidents may also be a good source of information.

To a large extent, performance-influencing factors comprise the interaction between people, technology and organization. People must be enabled to utilize their abilities to identify, prevent or handle failure, hazard and accident situations. Workplaces and systems must be designed with regard to man–machine interfaces, information presentation on screens and alarm management. Furthermore, it is important to ensure sufficient capacity and competence in the organization to avoid excessive time pressure and stress. Personnel must be familiar with checklists and procedures, which must be useful for the actual work tasks and incident situations. Checklists and procedures must also be easily accessible and comprehensible. Good management, workplace cooperation and physical working environment factors are also important.

To better understand the overall barrier concepts, Fig. 7.3 illustrates an example of technical, organizational and operational barrier elements involved in the barrier function "Detection of the kick during a drilling operation". Some examples of the corresponding performance requirements and performance influencing factors are also illustrated.

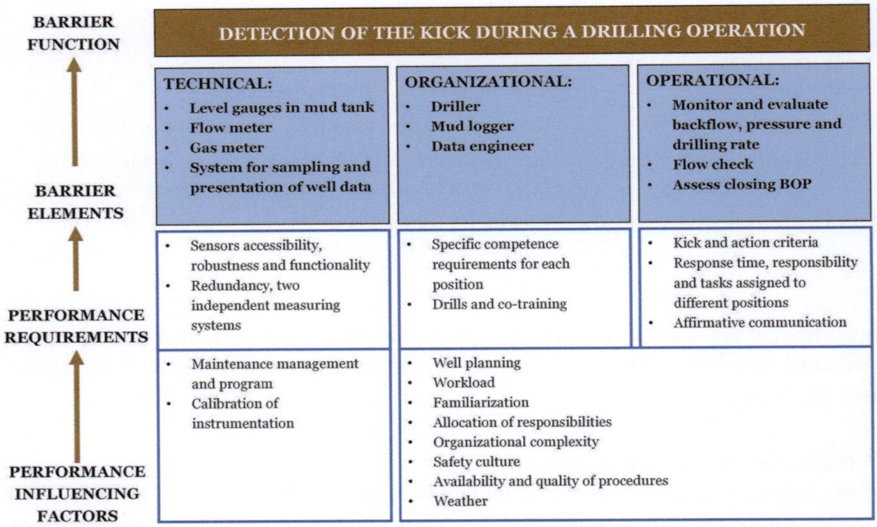

Fig. 7.3 Example of realizing the barrier function. Reprinted from Barrier memorandum 2017, with permission from Havtil

7.1.2 What Is Barrier Management?

Barrier management involves coordinated activities for establishing and maintaining barriers so that they fulfill their functions at all times. Barrier management represents an integral part of the enterprises' risk management, which in turn forms an integrated part of their corporate governance, as illustrated in Fig. 7.4.

Barrier management is important in order to have a clear process for handling failure, hazard and accident situations. Barriers must be established when the risk management process detects failure, hazard and accident situations in which there is a need for extra protection.

The primary purpose of barrier management is to establish and maintain barriers so as to be able to handle the risk faced at any given time. This is done by having barriers whose function is to help prevent failure, hazard and accident situations occurring or to restrict the consequences if they do occur. Thus, the aim of barrier management is risk reduction.

The barrier strategy refers to a plan for how barriers function, on the basis of the risk picture, are implemented in order to reduce risk. The barrier strategy includes establishing which barrier functions are necessary. It also determines which technical, operational and organizational barrier elements must be included in order to fulfil the function. Performance requirements for these elements must also be part of the strategy. Barriers must be established based on the specific risk picture for the individual areas of the facility or plant and for different activities or operations.

The principal characteristics of barrier management are illustrated in Fig. 7.5.

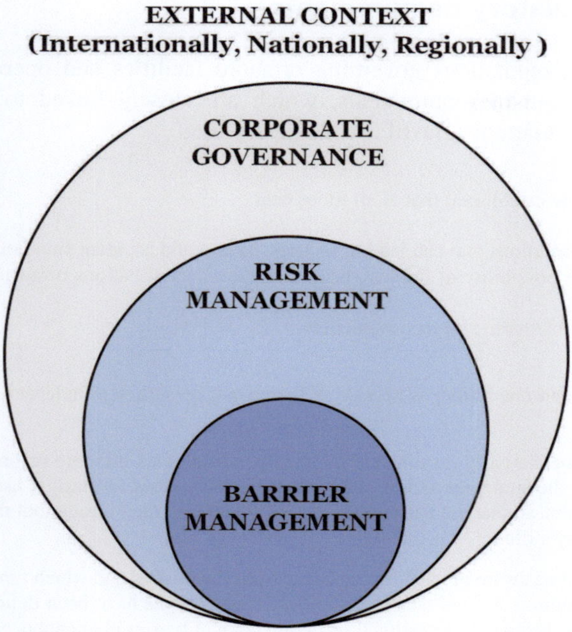

Fig. 7.4 Barrier management—an integral part of risk management and corporate governance. Reprinted from Barrier memorandum 2017, with permission from Havtil

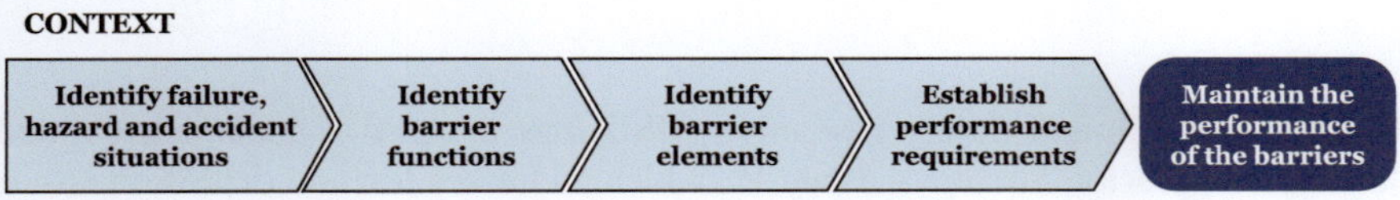

Fig. 7.5 Principal characteristics of barrier management. Reprinted from Barrier memorandum 2017, with permission from Havtil

7.1.3 What Is the Rationale Behind Barrier Management?

There is a truth, which should be realized, that no matter how safely and robustly the offshore petroleum activities are designed and operated, failure, hazard and accident situations can still occur. This is why barriers are needed and must then fulfil their functions in order to help handle such situations.

The detailed functions are to identify conditions that can lead to failure, hazard and accident situations at an early stage, reduce the possibility of failure and hazard and accident situations developing, limit possible harm and disruption. This is illustrated in Fig. 7.2.

7.2 Regulatory Requirements

The Norwegian regulations governing offshore facilities and operations encompass a wide range of requirements, which are closely linked to barriers. The management regulations (Havtil 2020) specify that:

> Barriers shall be established that at all times can:
>
> (a) identify conditions that can lead to failures, hazard and accident situations,
> (b) reduce the possibility of failures, hazard and accident situations occurring and developing,
> (c) limit possible harm and inconveniences.
>
> Where more than one barrier is necessary, there shall be sufficient independence between barriers.
>
> The operator or the party responsible for the operation of an offshore or onshore facility shall stipulate the strategies and principles that form the basis for design, use and maintenance of barriers, so that the barriers' functions are safeguarded throughout the offshore or onshore facility's life.
>
> Personnel shall be aware of what barriers have been established and which function they are intended to fulfil, as well as what performance requirements have been defined in respect of the concrete technical, operational or organizational barrier elements necessary for the individual barrier to be effective.
>
> Personnel shall be aware of which barriers and barrier elements are not functioning or have been impaired.
>
> Necessary measures shall be implemented to remedy or compensate for missing or impaired barriers.

The UK's Health and Safety Executive (HSE) has released a document entitled "Fire and Explosion Strategy—Issue 1", which addresses fire and explosion issues on offshore installations. This document provides a comprehensive overview of the existing knowledge concerning fire and explosion hazards, associated with their prevention, control, and mitigation. It identifies current areas of uncertainty and delineates the strategy areas where significant uncertainties necessitate further clarification. The focus of this document is on individual strategies and functional requirements for the prevention, control, and mitigation of fires and explosions, rather than on the management process.

The international standard ISO 13702 (ISO 2015) mandates the establishment of two strategies: the fire and explosion strategy (FES) and the emergency response strategy (ERS). Havtil references the ISO 13702 standard in its regulatory guidelines. The requirement for a barrier strategy can be viewed as a generalization of both the FES and ERS. These strategies are expected to articulate the roles and primary functional requirements of each system necessary for managing potential hazardous events within the installations.

7.3 Principles for Barrier Management

This section presents the principles for barrier management, covering specific assessment, prioritization of barriers, the interaction between different barrier elements, the role of maintenance within barrier management, and barrier management in all phases. As the preceding section titled "What Are Barriers?" has already addressed the principles in relation to barrier elements, performance requirements, and performance influencing factors, these principles will not be revisited herein to prevent unnecessary redundancy.

7.3.1 Specific Assessment for Barriers

The risk picture is an understanding and overview of potential failure, hazard and accident situations and how to protect against them. It is important to note that the risk picture will be different for different facilities and plants, areas and operations. An overall assessment of failure, hazard and accident situations and barriers at the facility and plant level will be important for platform and onshore management. For personnel working in a specific area or on a specific task, assessments are required for the specific areas of the facility and for the specific operations in order to obtain a good risk picture. This means that failure, hazard and accident situations must also be assessed in sufficient detail so as to identify the situations where the individual barrier element has a function.

In summary, effective barrier management must be grounded in a specific risk picture for the plant, facility, area or operation.

7.3.2 Prioritization of Barrier Functions and Elements

As a fundamental principle, priority should be given to likelihood-reducing measures over consequence-reducing measures, but there is often a need for both types of measures.

Barriers are normally categorized as passive or active. Passive barriers are present without needing activation or intervention, such as a firewall. Active barriers require activation or intervention. Normally, passive barriers should be preferred over active ones. Active barriers that are activated automatically are normally preferable to active barriers that require intervention and actions in order to be implemented. As a result, technical barrier elements are often preferred over elements that require human intervention. An assessment should be made of the reliability, strengths and weaknesses of the technical elements in relation to the organizational and operational elements in handling the specific failure, hazard and accident situation.

The strengths of technical systems are that they are "always" in place and ready to perform their tasks. Their strength is also evident when many signals have to be processed quickly and fixed actions must be performed in sequence or simultaneously. They are also designed to be reliable and robust and maintain their function and performance during necessary maintenance. Technical systems may, however, have weaknesses in terms of flexibility and their ability to handle unforeseen situations or combinations of situations. Increased complexity in the maintenance of the technical barrier elements may also introduce an increased risk of system failures occurring. Technical systems may also fail if non-predefined situations arise.

The strength of humans resides in greater flexibility, the ability to see new patterns, and assess complex and unforeseen situations ("black swan events"). However, humans have their weaknesses in that they may misunderstand complex situations, make errors, not be situationally aware and act slowly.

Therefore, it is often beneficial to distribute tasks between technical systems and people, as each has its own strengths and weaknesses. The boundaries between which functions are allocated to the technical systems and which to humans have changed over the years and will also change going forward.

7.3.3 Interaction Between Technical, Organizational and Operational Barrier Elements

Barrier functions are provided in differing degrees through the interaction between technical, organizational and operational barrier elements. In some cases, a barrier function could be provided purely by technical barrier elements.

> For example, gas detection in an area will cause automatic decoupling of ignition sources and shutdown of production. For the further handling of the leak, the control room operator can initiate manual pressure relief. This barrier function includes technical, organizational and operational barrier elements. The technical elements are the pressure relief system (from the panel in the control room, pressure relief valves, the flare drum to the flare stack). The organizational elements are the control room operator, and the operational elements are the actions which the control room operator performs in order to initiate pressure relief.

In the vast majority of barrier functions, technical barrier elements will be included to one degree or another. However, there are certain barrier functions in which personnel must perform various actions in order for the function to be provided.

Figure 7.6 illustrates the interaction between the technical, organizational, and operational barrier elements involved in realizing a barrier function. Furthermore, examples are provided of performance requirements and performance-influencing factors for the different barrier elements. The question "Who does what with which equipment in failure, hazard and accident situations?" encapsulates the interaction between the technical, organizational and operational barrier elements. In this context, "who" refers to the organizational barrier element, "what" denotes the operational barrier element and "which equipment" refers to the technical barrier element. This formulation also clarifies what is necessary for realizing a barrier function. Nonetheless, in practice, there has been a predominant emphasis

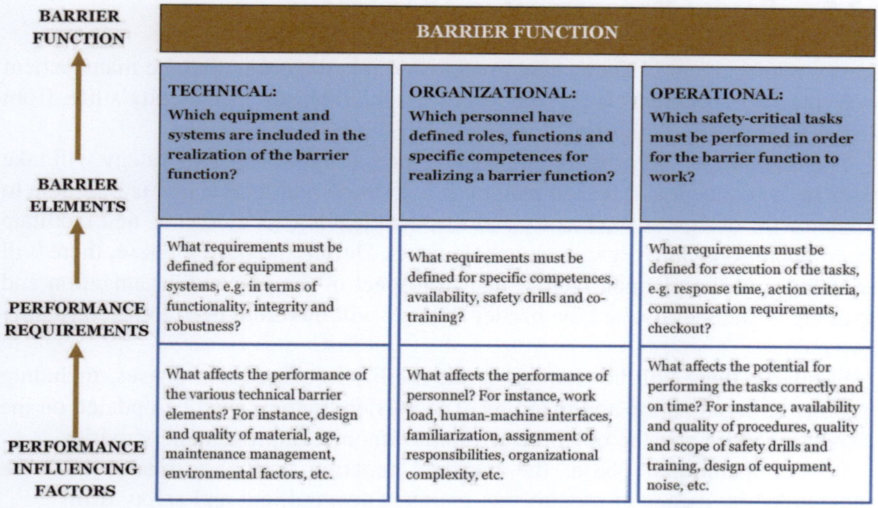

Fig. 7.6 Illustration of how a barrier function can be ensured. Reprinted from Barrier memorandum 2017, with permission from Havtil

on the technical barrier elements, while the organizational and operational barrier elements have received comparatively less attention and emphasis.

7.3.4 The Place of Maintenance in Barrier Management

Maintenance consists of verifying, maintaining and restoring the performance of barriers. Maintenance makes three core contributions to safety and barrier management:

- Maintenance helps maintain and, where appropriate, restore the technical aspect of the safe and robust solution so that failure, hazard and accident situations do not arise. This contribution is achieved by preventive maintenance (PM) or corrective maintenance (CM).
- Maintenance contributes to maintaining and, where appropriate, restoring the performance of technical barrier elements. This contribution is also achieved by PM or CM.
- Maintenance contributes to verifying the performance of the technical aspects of a safe and robust solution and the technical barrier elements. This contribution is normally achieved by functional tests or condition monitoring on barrier elements.

7.3.5 Barrier Management in All Phases

Barrier management is a persistent process and it is important that the management loop (plan-do-check-act) is continuously used in all phases of a facility's life, from design to removal, as illustrated in Fig. 7.7.

The foundation and primary emphasis in preparing the barrier strategy will take place in the concept and design phases. It is in these phases that one is best able to influence the design of the facility and the technical barrier elements, and facilitate sound operational and organizational elements. During the design phase, there will be changes to the planned facility and in respect of the planned organization and manning of the facility, and the barrier strategy will therefore need to be constantly corrected.

Similarly, there will be changes to the facility in subsequent phases, including in terms of organization and manning. The barrier strategy must be updated on the basis of changes and the outcome of monitoring and improvement activities.

In the operational phase, the barriers' function and performance must be safeguarded by means of monitoring, maintenance, training and safety drills.

7.4 Barrier Management Process

Havtil's model for the barrier management process is illustrated in Fig. 7.8, and consists of the following steps:

- Establishing a context for barrier management.
- Establishing the risk picture and barriers, including facility-specific and plant-specific barrier strategies and performance requirements.
- Adhering to and maintaining the barrier strategy in further activities, including:
  - operating in compliance with assumptions and requirements;
  - keeping the barriers intact and handling changes in the risk picture and barrier status when necessary.
- Monitoring and reviewing compliance of the barrier strategy and maintenance of barriers, including:

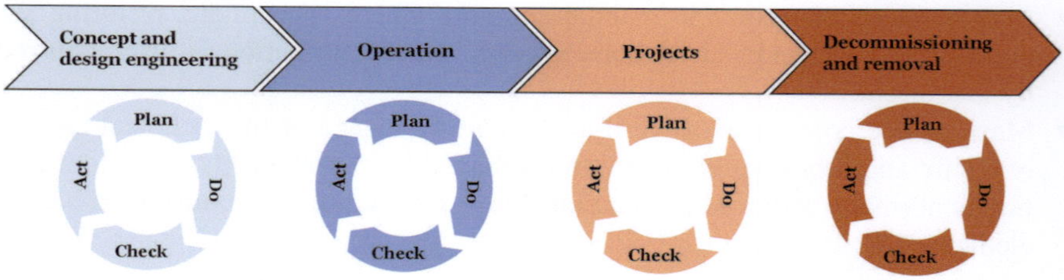

Fig. 7.7 Barrier management in life cycle phases. Reprinted from Barrier memorandum 2017, with permission from Havtil

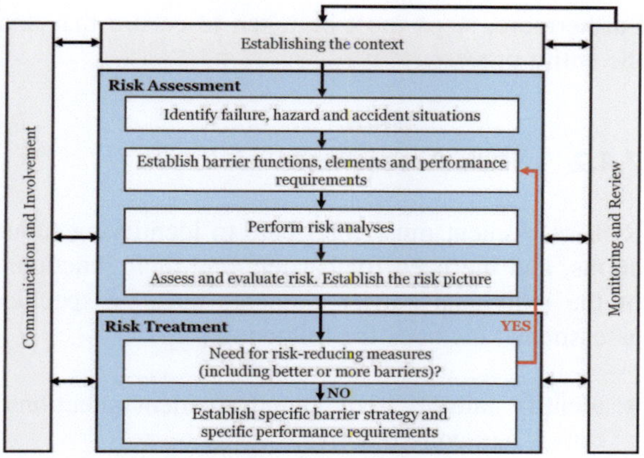

Fig. 7.8 The barrier management process during the planning phase. Reprinted from Barrier memorandum 2017, with permission from Havtil

  - collecting measurement data, results from safety drills and verifications in order to create the necessary foundation for this compliance.
- Establishing a basis for decisions on improvements where necessary, based on results from monitoring and reviews.

This process is developed on the basis of regulatory requirements and ISO 31000 standard (ISO 2018), which was introduced in the Second Lesson of this book. It can be seen that the upper part of the model aligns with the risk assessment and risk treatment framework outlined in the ISO 31000 standard, while the lowest step is tailored specifically for barrier management. This process will yield two primary outcomes: a specific barrier strategy and specific performance requirements.

7.4.1 Establishing the Context

The context is the interrelationships, operating parameters and guidelines (external, internal and project/activity-specific), which are relevant for performing the other steps in the barrier management process. In other words, it encompasses everything that could be directly or indirectly important for implementing and for shaping the strategy that is ultimately adopted. Factors of importance might include:

- requirements and guidelines in regulations, standards and enterprise-specific policies;
- enterprise-specific strategies, goals and principles for risk and barrier management;
- actual design, condition and location.

To achieve continuous improvement, which contributes to good and robust solutions, the requirement for risk reduction must also form part of the context.

Furthermore, steps must be taken to ensure that this is taken care of throughout the entire process.

7.4.2 Risk Assessment

Risk assessment must contribute to identifying failure, hazard, and accident situations, and the need for barriers and their functions. Furthermore, the properties of the individual barrier elements must be specified. The collective term "risk assessment" includes the following steps:

- Identify failure, hazard, and accident situations to determine their possible causes and consequences.
  - Identification of potential failure, hazard and accident situations must be performed in sufficient detail and specifically for the facility, plant, area and operation, so as to identify the situations that the individual barrier element has a role in handling. Realistic combinations of failure, hazard and accident situations must be assessed, since such combinations may determine the dimensions of barriers.
- Establish barrier functions, barrier elements, and associated performance requirements and performance-influencing factors.
  - Once potential failure, hazard and accident situations have been identified and assessed, the necessary barrier functions and barrier elements must be identified and the process of establishing associated performance requirements and mapping performance-influencing factors should start.
- Carry out risk analyses and necessary safety studies.
  - Risk analyses conducted as part of decision support for barrier management must be planned and performed in such a way that they are sufficiently detailed and appropriate for their use.
  - Assumptions on which a decision is based must be expressed so that they can be followed up. The specific barrier strategy should indicate which assumptions are significant for the individual barrier function and the individual barrier element. This does not necessarily mean that all assumptions have to be described in one place, but it must be easy to obtain such information by referring to the barrier strategy.
- Assess and evaluate risk, including sensitivity and uncertainty-establish the risk picture.
  - According to the aforementioned definition of risk picture, it means relevant, usable and appropriate information for the individual in the specific situation, including the significance of their own choices, decisions and actions.
  - Sensitivity and uncertainty assessments must be performed as part of a risk analysis. The primary purpose of this is to give users of the analysis the best possible basis for understanding its strengths, weaknesses and limitations, as well as which suppositions, assumptions or assessments are of importance for the results of the analysis. It is also important to understand the

uncertainty underlying the input parameters and assessments on which the analysis is based. Sensitivity and uncertainty must be assessed and used in the work of risk handling, whether that involves communicating the need for risk-reducing measures in barrier strategies or detailing specific performance requirements.

– In most cases, it is not sufficient to exclusively use quantitative risk analyses for deciding if the individual barriers are needed. Nor is it sufficient to establish specific performance requirements for the individual barrier element.

7.4.3 Risk Treatment

Risk must be reduced as far as practicable. There is a much greater chance of achieving good solutions by working in a structured and targeted way to reduce risk in the concept and design phases. As a rule, this will also yield the best solutions in terms of execution, cost and operation.

The requirement to reduce risk is not confined to measures that can be quantified in the results of a risk analysis. For example, "preventing hot surfaces on equipment which could be exposed to diesel oil leaks" is a sensible measure in most cases, even if its effect cannot be quantified in a QRA. Common sense should therefore be the guiding principle when assessing the effect of such measures, rather than relying exclusively on the results of risk analysis and cost-benefit analysis. On the other hand, risk analysis will in many cases represent a necessary and important supplement when assessing the effect of various measures. In other words, it is a case of finding the tools, analyses and so forth which provide relevant decision support for the various issues.

Establishing a Specific Barrier Strategy

As noted before, a specific barrier strategy is a plan for how barrier functions, based on the risk picture, are to be implemented to reduce risk. The definition emphasizes that the starting point of a specific barrier strategy is the risk picture, which will normally be obtained on the basis of a QRA study; however, alternative approaches may also be applicable. It should be noted that QRA studies have faced considerable criticism due to their predominant focus on FAR or AIR values. Thus, it is imperative that risk contributions are expressed using additional risk parameters beyond just FAR or AIR values. Within the context of the Norwegian regulatory framework, additional risk parameters may include the frequencies associated with the impairment of the main safety functions, which encompass escape ways, safety areas, main load carrying capacity, etc.

The basis of a barrier strategy ought to be established upon the following principles:

- It should be designed so that it helps the parties involved gain a common understanding of the basis of the requirements for the different barriers, including:

  - which phases, operations and activities the strategy has been established for;
  - which failure, hazard and accident situations may occur in the phases, operations and activities the strategy has been established for;
  - which barrier functions are required to handle these situations;
  - where additional information is to be found for the performance requirements that specifically apply to the individual barrier.
- It should be sufficiently fine-grained for the individual plant (e.g., area, system, equipment) and phases, operations and activities.
- It should be kept updated at all times.
- It should identify which roles or tasks the different barrier functions have.
- It should identify important assumptions which are significant for the individual barrier function and the individual barrier element.
- It should identify the relationship between strategy and performance requirements established for the individual barrier. The strategy should provide information about where the different performance requirements for the individual barrier element and the individual barrier function are described.

A large offshore installation normally consists of a limited number of main areas (e.g., living quarters, utility area, drilling and wellhead area, process area, and storage area for hydrocarbons), which nevertheless can be sub-divided into many sub-areas and rooms that require special protection measures. It is advisable that the establishment of a specific barrier strategy follows the separation of main areas, rather than being applied to the individual sub-area or room.

Establishing Specific Performance Requirements

Establishing and using a specific barrier strategy and associated performance requirements will be crucial for the ability to establish effective barrier management. Good usage requires that this documentation is sufficiently well-known to relevant personnel in the individual facility.

It is essential to structure barrier functions, barrier sub-functions and barrier elements systematically. This could be achieved by a barrier diagram, as shown in Fig. 7.9. The performance requirements are primarily set for barrier elements, but in certain cases also for barrier systems and barrier functions or subfunctions. For instance, performance requirements set for "ship collision barrier" and "containment barrier" are primarily focused on the specific barrier functions of "preventing ship collisions" and "preventing leaks," rather than on the individual barrier elements.

It may be beneficial to group the established performance requirements into performance standards at the system or function level, as has been implemented by some enterprises. The NORSOK S-001 standard (Standards Norway 2018) is built around this thinking. In addition to specifying performance requirements for barrier elements, a performance standard should also clarify the interfaces with other barriers.

The establishment of performance requirements encompasses several key principles, which are outlined as follows:

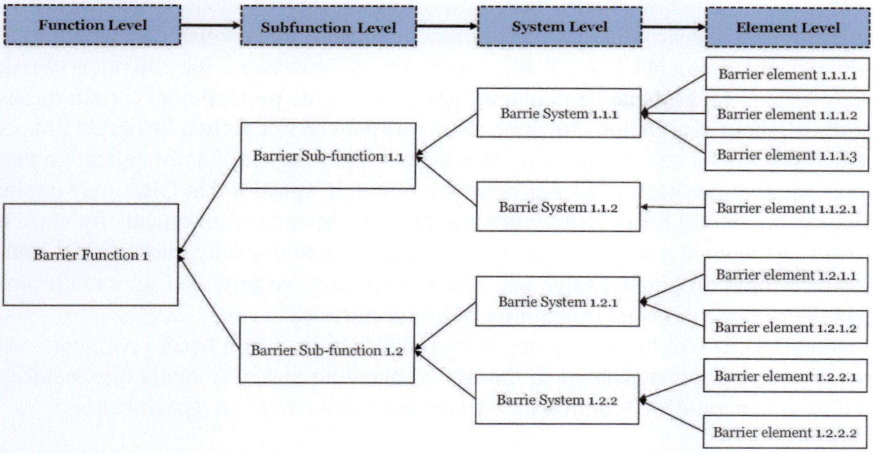

Fig. 7.9 Illustrative barrier diagram

- The relationship between performance requirements for barrier elements and failure, hazard and accident situations should be explicitly clarified.
- The relationship between strategies and barriers should be clarified.
- The maintenance plan and execution procedures designed to safeguard the performance of barriers in every phase of the life cycle should be clearly stated.
- Performance requirements, along with their corresponding rationales, should be well understood by the relevant personnel on the individual installation.

As illustrated in Fig. 7.8, the initial step in establishing performance requirements is the identification of hazards, which has been advocated by Havtil for increased emphasis. Furthermore, it is crucial that risk analyses are conducted with adequate detail to serve as a robust foundation for establishing specific performance requirements.

During the process of performing risk analyses, it is common to have many assumptions and premises. If these assumptions and premises have possible influences on the barrier performance, they should be clearly defined as performance requirements.

It is crucial to establish, present and evaluate risks based on their actual purposes. The risk picture needs to be established to determine the necessary barrier functions and to ensure that barrier elements possess the required performance. Additionally, sensitivity studies and uncertainty evaluations play a crucial role in establishing performance requirements. Their primary goal is to provide users with a solid foundation for comprehending the strengths, weaknesses and limitations of the analyses, along with the key assumptions, premises, or evaluations necessary for appraising the outcomes.

It is important to recognize that, in most cases, a QRA alone is insufficient to determine the necessity of specific barrier elements or to establish the specific performance requirements. For instance, in certain circumstances, the outcomes of risk analyses may suggest that particular barriers, such as fire protection of certain rooms on the offshore installation, are unnecessary as they do not yield a favorable impact on the risk assessment results. This should not be misconstrued as an indication that the room does not need to be fire-resistant; rather, it signifies that QRA may not be an appropriate tool for providing input to the performance requirements for certain barrier elements in this context. Actually, it may be mandated by international standards or common practices that fire protection should be provided for such rooms regardless of the assessed probability of fire occurrence.

Moreover, it will be misleading to imply that decisions on barrier elements rest merely with barrier management during the planning phase. Actually, the majority of barrier elements are required by the standards and enterprise specifications.

Communication and Involvement

It must be ensured that the risk picture and barrier strategy and associated performance requirements are communicated to and anchored with both internal and external stakeholders, and are appropriate throughout the entire barrier management process. This is intended in part to ensure:

- good quality-by drawing on relevant expertise and experience throughout the process, including when establishing the context, when conducting risk assessments and management, and for supervising and monitoring at all times;
- participation by, and a sense of ownership among, stakeholders who will be affected by decisions in every phase;
- understanding of the background to decisions;
- that risk analyses are communicated in such a way that target groups obtain a nuanced and coherent presentation of the analysis and its results;
- that the barrier strategy is actively used to provide those involved with a common understanding of the risk picture and the basis for the requirements specified for the various barriers.

Communication and anchoring of the risk picture and barrier strategy are not to be regarded as an independent activity, but as one that will pervade the whole barrier management process in every phase.

7.5 Barrier Management in the Design and Operational Phases

This section discusses the critical factors to consider for effective barrier management in the design and operational phases.

7.5.1 Barrier Management in the Design Phase

Barrier management in the design phase of a new facility follows, in principle, the planning element in the barrier management loop.

The following factors are crucial to consider for effective barrier management during the design phase.

- Establishment of a barrier strategy is an iterative process.
- Barriers must be maintainable, while the performance must be verifiable and the operation must be achievable within the defined limitations.
- Manning for safeguarding barrier functions and operational tasks should be sufficient.
- Barrier management is an interdisciplinary activity.

7.5.2 Barrier Management in the Operational Phase

To fulfill the requirements pertaining to barriers, as specified during the design phase, it is essential to employ and follow up the barrier strategies during operation, as well as implement improvements in order to reduce risk as the need arises.

Barrier management has faced significant challenges in effectively linking the design phase with the operational phase. The foundation for ensuring safe operations is established during the design phase. On this basis, by actively following up the barrier elements and performance requirements during operation, a close connection between these two phases can be achieved, potentially bridging the gap often identified during audits. However, practical experience indicates that the assumptions made during the design phase can be difficult to implement in the operational phase. Additionally, it is known that assumptions and parameters may evolve during operations due to changes in operating conditions, modifications, and other related factors. Through ongoing operational experience, some initial assumptions may become invalid. Moreover, some aging facilities currently in operation have lacked a systematic basis for effective barrier management from their concept and design phases. These cases highlight the necessity of having a structured framework for barrier management, as well as ongoing enhancements throughout the entire operational phase.

The following factors are crucial to consider for effective barrier management during the operational phase.

- Operating and using the facility in compliance with assumptions, requirements and technical conditions.
  - What is the condition of the barriers? Do they meet the requirement? Do they fulfil their role?
  - Compensatory measures must be implemented to rectify missing or impaired barriers.

- Ensuring and maintaining the necessary barrier performance.
  - Technical barrier elements must be tested, inspected and maintained.
  - Organizational and operational barrier elements must be trained and drilled.
  - Performance requirements must be reviewed and updated in the operational phase.
  - The effect of modifications and changes must be assessed. Organizational changes are also considered changes.
- Measurement and verification of the barriers' performance.
  - The monitoring, testing and verification of operations should ensure compliance with the established prerequisites.
  - Non-conformities that affect the risk picture and barriers must be assessed and dealt with together.
  - Measurement and verification of performance can be usefully based on existing systems, such as evaluation of training and safety drills.
  - Results of measurements and verifications must be communicated to those who have tasks in a barrier function.
- Maintaining control of performance-influencing factors.
  - Maintenance and retaining competence are important performance influencing factors for managing changes, along with good decisions based on sound risk assessment through cooperation between management, operational personnel and safety delegates.
  - Management is a performance influencing factor, ensuring systems and resources are in place, so that barriers are established and their integrity is assured.
  - Good risk understanding is a key performance influencing factor to achieve the compliance with procedures, working practices and follow-up.
- Following up and enhancing the barrier management system.
  - Follow-up must identify needs for improvements in the barrier management process and implement necessary improvement measures.

Establishment of the foundation for safeguarding these conditions will play a key role in the importance of barrier management for the preparations of the operation. Securing effective communication and involvement in the planning stage will be important for establishing and following up on this foundation. Figure 7.10 gives examples of the above-discussed factors, which are key to operation.

7.6 Structure of Barriers and Hazards

This section suggests how the connections among hazards, main areas requiring special protection, and barriers are structured on the offshore petroleum installations.

Figure 7.11 presents a structure principle of typical barrier functions and hazards for a combined offshore drilling and production installation. The color coding

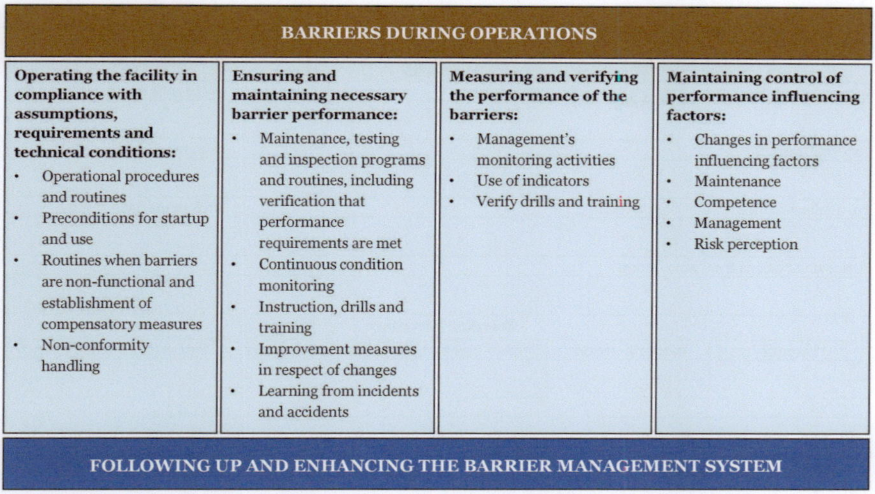

Fig. 7.10 Examples of factors that are key to barrier management during operations. Reprinted from Barrier memorandum 2017, with permission from Havtil

employed differentiates the connections among the barrier functions and the protection of life (BF-10), environment (BF-11), safety–critical rooms (BF-12), as well as main load-bearing structures (BF-13). The connections between hazards and barrier functions indicate how different barrier functions deal with different hazards.

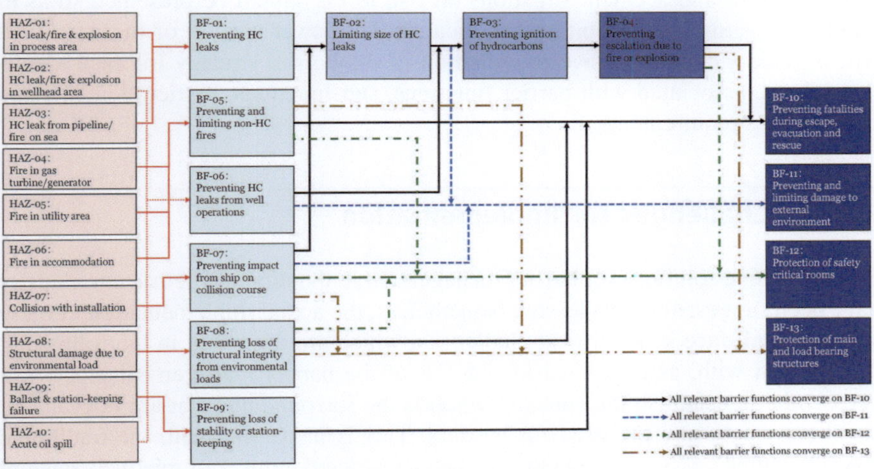

Fig. 7.11 Structure principle of typical barrier functions and hazards. *Source* Adapted from Vinnem and Røed (2020)

<table>
<tr><td colspan="5">Installation:</td></tr>
<tr><td colspan="5">Area/Room:</td></tr>
<tr><td colspan="5">Hazards</td></tr>
<tr><td colspan="4">Internal hazards:</td><td>Reference doc.</td></tr>
<tr><td colspan="4"></td><td></td></tr>
<tr><td colspan="4">External hazards:</td><td>Reference doc.</td></tr>
<tr><td colspan="4"></td><td></td></tr>
<tr><td colspan="4">Particular hazards:</td><td>Reference doc.</td></tr>
<tr><td colspan="4"></td><td></td></tr>
<tr><td colspan="5">Risk contributions from area/room:</td></tr>
<tr><td colspan="5"></td></tr>
<tr><td colspan="5">Barrier Strategy</td></tr>
<tr><td>Hazards</td><td>Barrier functions</td><td>Sub-functions</td><td>Barrier elements</td><td>Performance requirements</td></tr>
<tr><td></td><td></td><td></td><td></td><td></td></tr>
<tr><td></td><td></td><td></td><td></td><td></td></tr>
<tr><td></td><td></td><td></td><td></td><td></td></tr>
<tr><td></td><td></td><td></td><td></td><td></td></tr>
<tr><td></td><td></td><td></td><td></td><td></td></tr>
</table>

Fig. 7.12 Suggested format for the presentation of a specific barrier strategy

On this basis, the specific barrier strategy should be further designed and described to express these connections between hazards, main areas with special protection needs, and barriers. A possible format is suggested for the barrier strategy, as shown in Fig. 7.12. The upper section of the document is designated for the identification of the installation and area/room. Hazards are categorized into two distinct types: internal hazards and external hazards. Further, it is essential to focus on any particular hazards pertinent to the installation, as these hazards may necessitate supplementary barrier functions or elements. Then, the effects of failure, hazard and accident situations on risk levels should be presented so as to pinpoint hazards that exhibit high criticality. The lower section of the document will provide a comprehensive presentation of the barrier strategy for each identified hazard, associated with barrier functions, sub-functions, barrier elements and performance requirements.

7.7 Challenges for Implementation

Some tentative solutions for barrier management in the offshore petroleum industry have been suggested in this lesson. Nonetheless, the actual implementation of these solutions will face some critical challenges, which are discussed in the following.

To begin with, as illustrated in Fig. 7.8 of the barrier management process, it is essential to establish the context, which is the surroundings that are relevant for assessing safety and the need for barriers. This is in parallel with the traditional HAZOP studies. It is advisable to establish a restricted number of relatively general contexts, avoiding excessive details.

Secondly, the structure principle of barrier functions and the corresponding hazards is a challenge. The proposed structure in Fig. 7.11 is supposed to be a general structure and the details in the hazard categories will vary with the installation type, the process media, etc. Surely, there exist other ways to structure the hazards and what calls for special attention is that the hazard categories should be kept at a low level so as to enable easy communication.

Thirdly, it is challenging for the structure of barrier functions, barrier elements and performance requirements. In practice, some offshore petroleum enterprises prefer to structure the strategies for barrier systems in place of barrier functions, such as the gas detection system, fire detection system, emergency shutdown, etc. In this situation, of particular note is the relationship between the barrier systems and barrier functions. The following questions need to be addressed: Which barrier systems are involved in supporting each barrier function? What is the overall performance for each barrier function?

Further, the most important challenge for barrier management is how to distinguish between what needs to be tailored and what can follow the industry practice in the establishment of barrier strategies. The philosophy document (Havtil 2017) makes an impression that a specific barrier strategy should always be tailored to the facility concept and its context. Actually, the vast majority (probably more than 90%) of barrier systems and requirements on an installation can follow the common practice. This distinction must be taken seriously for the establishment of barrier strategies so as to avoid simply copying from a similar installation without any thought.

Lastly, Havtil has realized the key importance of organizational and operational barriers. However, their implementation effects in practice are still unclear.

7.8 Questions

Q1: Please explain the relationship between risk analysis and barrier management.

Q2: In what ways can operational barrier elements be effectively monitored?

Q3: Regarding the hazard pertaining to "Passing ship on collision course with installation", please specify possible barrier strategy, barrier functions (also sub-functions if needed) and barrier elements.

Q4: If the maintenance of barrier elements falls behind the planned schedule, what effects might this have on the barrier management?

Q5: Why is it important to communicate barrier strategy and barrier elements status to operational personnel?

References

Ersdal G (2014) Safety of structures. Compendium at University of Stavanger

Havtil (2013) Principles for barrier management in the petroleum industry. Stavanger, Norway

Havtil (2017) Barrier memorandum: principles for barrier management in the petroleum industry. Stavanger, Norway
Havtil (2020) Regulations relating to management and the duty to provide information in the petroleum activities and at certain onshore facilities. Last amended 16 Dec 2020
ISO (2015) Control and mitigation of fires and explosions on offshore production installations—requirements and guidelines. ISO 13702. International Standards Organization, Geneva
ISO (2018) Risk management—guidelines. ISO 31000. International Standards Organization, Geneva
Standards Norway (2018) Technical safety. NORSOK S-001. 6th edn
Vinnem JE, Røed S (2020) Offshore risk assessment: principles, modelling, and applications, 4th edn. Springer, London

Eighth Lesson: Indicators for Major Accident Risk

8

If you cannot measure it, you cannot improve it.

—*Lord Kelvin (1883)*

Abstract

Conventional wisdom has long assumed that the major accident risk directly correlated with occupational accident statistics. Yet, the facts tell a different story. This lesson focuses on major accident risk indicators that could maintain high levels of awareness, motivation and emphasis on prevention against potential major accidents. The lesson begins by clarifying indicator concepts and definitions. In response to the conceptual confusion surrounding indicator concepts, the authors advocate adopting the terms of precursor indicator and barrier indicator. Then, the lesson explores the need for major accident risk indicators, emphasizing their particular characteristics. After that, a comprehensive set of evaluation criteria for major accident risk indicators is proposed and discussed to avoid the "garbage in-garbage out" problem. Finally, major accident risk indicators for the nation and enterprise are recommended. The lesson equips readers with the theoretical foundation as well as practices in the development and application of major accident risk indicators.

8.1 Indicator Concepts and Definitions

Indicators typically refer to numerically measurable variables, which serve to characterize the condition, level or status of a broader phenomenon or aspect of reality. In the first place, an important distinction should be made between indicators associated with safety and risk (Øien et al. 2011). Risk indicators are derived

X. Zhen and J.-E. Vinnem, *Ten Lessons on Risks*,
https://doi.org/10.1007/978-3-031-99064-9_8

from the risk model, whereas safety indicators are developed on the basis of the safety performance. Viewed in this light, the confusion of concepts between "safety indicator" and "risk indicator" could be significantly minimized.

It is worth noting that a debate regarding the indicator concepts and definitions often occurs when it comes to the differentiation between leading and lagging indicators, as evidenced by a special issue in the *Journal of Safety Science* (Hale 2009). Hopkins (2009) identifies two primary perspectives on this distinction. The first perspective is the relativistic approach (HSE 2006; Erikson 2009; Dyreborg 2009), which posits that a specific indicator may be considered leading in relation to one objective, yet lagging in relation to another. Nonetheless, this relativistic perspective struggles to consistently maintain a clear distinction between the two types of indicators. The second perspective represents the absolutist approach (Hale 2009; Hudson 2009), which is grounded in the bow-tie model, as shown in Fig. 1.2. This approach asserts that indicators located to the left of the "knot" are considered leading, whereas those situated to the right are classified as lagging. Nevertheless, ambiguity persists due to the uncertainty surrounding the exact location of the "knot" within the sequence. Ultimately, both the relativistic and absolutist approaches fail to provide a fully satisfactory explanation of the fundamental differences between leading and lagging indicators.

Kjellén (2009) defines a leading indicator as one that changes before the actual risk level has changed. This definition is supported by the authors. This focus on leading indicators is particularly critical within the framework of major accident risk, as major accidents and even precursor events are rare in high-reliability industries. Therefore, it is essential to maintain a sustained level of motivation and awareness that we possess indicators that change before the risk level has changed, so as to avoid the next major accident.

The authors would like to argue that an academic focus on the definitions and nuanced differences of indicators may be interesting, but not yield significant benefits. Furthermore, the authors would like to argue that the terminology surrounding leading and lagging indicators could be abandoned and replaced with the terms of precursor indicators and barrier indicators as follows (Vinnem 2010; Zhen et al. 2022).

- Precursor indicators based on occurrence of incidents and near-misses
- Barrier indicators based on performance of barriers installed to protect against major hazards and their potential consequences.

As has been stated by the definition of Kjellén (2009), precursor indicators are obviously lagging indicators, as the risk level has already changed, at least temporarily, when a precursor event occurs. In contrast, the barrier indicators, which reflect changes prior to the actual changes in the risk level, are leading indicators. The concepts of precursor indicators and barrier indicators have been successfully adopted and implemented in the Norwegian Risk Level Project (Havtil 2024), which will be further elaborated in the subsequent sections.

8.2 Need for Major Accident Risk Indicators

Prior to the year 1990, it had long been assumed that occupational accidents served as a primary metric for assessing major accident risks. A notable illustration of this belief is Heinrich's triangle (Heinrich 1931), as shown in Fig. 8.1. Consequently, considerable emphasis and efforts were placed on health, safety and environmental indicators associated with occupational injuries within the offshore petroleum sector, particularly concerning the number and frequency of lost-time injuries (LTI) (Visser 1995; Pasman and Rogers 2014; Pasman 2015). The focus on LTI indicators indicated that enhancements in major accident risks were directly correlated with improvements in LTI statistics.

Recent major offshore accidents, such as the Macondo well blowout (2010) in the US, the PL 19-3 oil spills (2011) and the PL 19-3 blowout and fire (2021) in China, have heightened awareness of the management distinction between major accidents and occupational accidents. In particular, investigations into these major accidents have revealed that the erroneous belief in a strong correlation between major accidents and occupational accidents played a crucial role in creating the conditions that enabled such major accidents to occur.

Therefore, it is imperative to identify what the particularities of the major accident risk and occupational accident risk are. Occupational accident risk implies no or very short time lag for feedback. For instance, if you do not wear personal protective equipment where it is mandatory, it is quite visible. Nonetheless, major accident risk is often characterized by very long delays between an error and its consequence. Consider one of Statoil's incidents in 2016 (a fire on Statfjord A platform). In this case, the failure to properly investigate an incident in 2003 resulted in the 2016 fire because the Preventive Maintenance (PM) program had not been

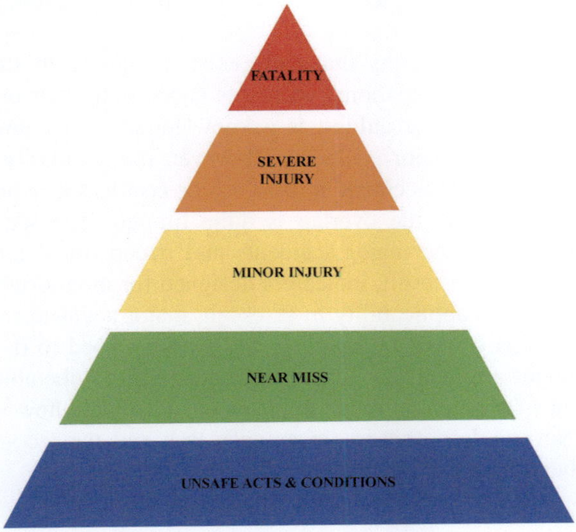

Fig. 8.1 An illustration of Heinrich's triangle

updated in 2003 due to inadequate incident investigation. Further, the rare occurrence, as a distinctive characteristic of the major accident, merits attention. On the NCS, there have been no major offshore accidents with fatalities since the year 1985. Actually, even precursor events for potential major accidents are pretty rare, occurring on the order of less than one event per installation annually. In addition, the particularity lies in the causal mechanisms of major accidents, which usually bear little resemblance to occupational accidents. For instance, hydrocarbon leaks on offshore petroleum installations can be taken as an example to illustrate this characteristic as follows.

On offshore petroleum installations, a common classification is made between hydrocarbon leaks with flow rates exceeding 0.1 kg/s and those below this threshold. Leaks with flow rates above 0.1 kg/s are relatively infrequent, whereas numerous minor leaks typically occur at rates below this level. Notably, only leaks exceeding 0.1 kg/s are regarded as precursors to major accident risks, as the resultant gas cloud from such leaks could potentially be large enough to trigger an explosion or fire. Some experts may argue that the frequency of leaks below 0.1 kg/s serves as a good indicator for the occurrence of more serious leaks. However, empirical evidence does not support this assertion. The underlying reasons for this discrepancy are analogous to those explaining why the LTI rate is not a good indicator of major accident risk. The causal mechanisms leading to major leaks (above 0.1 kg/s) differ fundamentally from those associated with minor leaks (below 0.1 kg/s). Minor leaks are often attributed to seepage around valve stems resulting from the brittle failure of rubber seals, which can be readily remedied through tightening or replacement of the seals. Importantly, these mechanisms rarely escalate into more serious leaks. In contrast, major leaks could result from valves being inadvertently left open during maintenance, unintentional disassembly of pressurized equipment or corrosion-induced perforations. Thus, the causal pathways for major accidents are fundamentally distinct from those of occupational accidents.

It is noteworthy that there exists a significant similarity between major accidents and occupational accidents concerning their organizational root causes. For instance, safety culture is acknowledged as a common cause for both occupational and major accidents. However, this similarity does not justify treating the occupational accident and the major accident as equivalent. Beyond this point of resemblance, the overlap is quite limited. The sequences, delays, and possible actions in the major accidents and occupational accidents are significantly different. As a result, indicators designed for monitoring occupational accidents have restricted applicability in assessing major accident risk.

This context underscores the pressing need to develop effective indicators for major accident risk. These indicators ought to be able to detect early warning signals for potential major accidents and measure how well safety is being managed. Nevertheless, this is not an easy task and the recognized silver bullets have not been found or invented.

The validity of the following general objectives of major accident risk indicators, as articulated by the author (Vinnem 2000), has been corroborated by over four decades of experience. These objectives include:

- Indicators should encompass a wide range of incidents, from minor near misses to severe and complex accident sequences.
- Indicators should reflect both the hazard potential and the actual outcomes of incidents.
- The priority of the incident in question should be reflected by indicators. For instance, unauthorized hot work conducted in a safety–critical area is of greater importance than the failure of a single fire water pump.
- There is a need for indicators that reflect the performance of individual systems and provide an overall assessment of safety levels in relation to potential major accidents.

8.3 Evaluation Criteria for Major Accident Risk Indicators

In the evaluation of practical applications or novel proposals for major accident risk indicators, appropriate evaluation criteria should be established in advance to avoid the "garbage in-garbage out" problem. This research topic is challenging, and the existing literature on this topic is limited. Notably, most of the established criteria are not specifically designed for major accident risk, making them applicable to all types of risk indicators.

A thorough examination of the existing works (Haugen et al. 2011; Kjellén 2000, 2009; Vinnem et al. 2003, 2006) has led the authors (Vinnem 2010; Zhen et al. 2019) to propose a comprehensive set of evaluation criteria for major accident risk indicators, which are outlined as follows:

(1) *Observable and quantifiable*:

Major accident risk indicators must be observable and quantifiably measurable through recognized data collection methods and measurement scales. Typically, these indicators are represented by a ratio scale, such as the number of hydrocarbon leaks (exceeding 0.1 kg/s) per million man-hours. In practice, ensuring reliable reporting can be challenging. For instance, the precise quantification of leaks on offshore installations is often difficult, and the data can be tainted due to external factors. To further exemplify this point, in instances where a hydrocarbon leak coincides with a personal injury, the primary focus tends to be directed towards the injury, often resulting in the hydrocarbon leak being neglected. Focusing data collection on events of medium and high severity can significantly enhance reporting reliability. Moreover, quantification is crucial as it is possible to measure the performance in a consistent manner.

(2) *Intuitiveness*:

What is measured by major accident risk indicators should be intuitively recognized by the workforce as important for preventing major accidents. For instance, the suggested major accident risk indicator, which is represented by the number of years spent in the same position, may not resonate intuitively with the experienced workforce. In addition, it is advisable that complex calculations are not required by major accident risk indicators. For instance, a straightforward relationship, such as a decrease in observations correlating with improvement, is more likely to maintain confidence in the data. Conversely, if highly complex calculations are necessitated, the confidence of this relationship may be lost.

(3) *Validity*:

The indicators must provide a valid status measurement of major accident risk, indicating that the indicators should accurately reflect what they are intended to measure. For example, indicators derived from the frequency of LTIs pertain to occupational accident risk and are therefore not valid for assessing major accident risk.

(4) *Regular monitoring*:

The status of major accident risk indicators ought to be capable of being monitored at regular intervals (e.g., monthly, quarterly, semiannually, etc.) to ensure that a sufficient volume of valid data is generated. In general, the number of observations (e.g., incidents, barrier failures, etc.) for the system being analyzed should amount to the order of a dozen within a given time frame, such as one year.

(5) *Sensitive to change and capable of indicating trends*:

Major accident risk indicators ought to be capable of exhibiting a high degree of sensitivity to changes, thereby enabling early warnings by capturing significant changes within a socio-technical system that could significantly affect major accident risk. For instance, the number of occupational accidents resulting in fatalities would normally be insensitive to change. An effective set of major accident risk indicators should reflect changes in risk levels and highlight areas for potential improvement.

(6) *Robust against manipulation*:

Indicators should be designed to prevent organizations from presenting a misleadingly favorable image to "look good" through changing the reporting behavior, rather than implementing fundamental changes that reduce the risk level.

(7) *Robust against conflicting campaign signals*:

Psychological and organizational reasons could result in too low reporting of major accident risk indicators. For instance, incentive structures that reward the absence of incidents may discourage reporting, as employees might interpret these incentives as favoring unreported incidents.

(8) *Combination of lagging and leading indicators*:

It is advisable to prioritize leading indicators over lagging indicators, as reporting on preventative measures is generally more motivating than reporting on the occurrence of near misses or incidents. Nevertheless, lagging indicators can serve as valuable warnings for potential major accidents in the future. Therefore, a dual approach utilizing both types of indicators is recommended, with a primary focus on leading indicators supplemented by lagging indicators.

(9) *Reflection of hazard-formation mechanisms*:

It is crucial that major accident risk indicators closely reflect hazard-formation mechanisms, thereby maintaining and enhancing awareness of the risk mechanisms. For instance, the frequency of hydrocarbon leaks serves as a pertinent major accident risk indicator, as such leaks represent one of the most severe accidents occurring on offshore installations.

8.4 Major Accident Risk Indicators for the Nation and Enterprise

For the purpose of enhancing the implementation of major accident risk indicators within engineering practices, it is important to determine the level (nation, enterprise, or installation) at which the precursor indicators and barrier indicators will function most effectively. This differentiation is vital to ensure that these indicators are properly tailored to the specific needs and requirements associated with different levels.

Typically, precursor indicators are only effective at the national level and are generally unsuitable for application at the installation or enterprise levels, with the exception of certain large offshore oil and gas enterprises, such as CNOOC in China and Equinor in Norway. This limitation arises from the infrequency of incidents, averaging typically less than one per installation annually. This leaves precursor indicators with insufficient data to provide meaningful feedback. Even so, it is advisable to employ precursor indicators at both the installation and enterprise levels to sustain focus and awareness regarding potential major accidents.

In contrast, barrier indicators are applicable across all levels, i.e., the installation level, the enterprise level and the national level. This is because the substantial data is available from barrier testing. Nonetheless, the emphasis should be shifted more towards the examination of the suitability of these barrier data.

This section will address major accident indicators at the national and enterprise levels, respectively.

8.4.1 Major Accident Risk Indicators at the National Level

In the year 1999, Havtil initiated the project titled "Trends in Risk Level in the Norwegian Petroleum Activity", which is known by its Norwegian acronym, RNNP. This project has been significant and unique as major accident risk indicators that are specifically tailored for the petroleum sector at a national level are proposed.

This section presents a structured overview of the RNNP project, focusing on precursor indicators and barrier indicators associated with major accident risk. Notably, the RNNP project encompasses a wide array of indicators for major accident risk, occupational accident risk, occupational diseases, work environment, risk perception and maintenance work. This section will focus on the major accident risk indicators.

Purpose, Objectives and Key Limitations

(1) *Purpose*

In the late 1990s, a dispute arose among the primary stakeholders in the Norwegian petroleum sector. The union representatives and regulatory authorities expressed significant concern regarding the perceived increase in risk levels associated with offshore petroleum operations. In contrast, the enterprise management and their representatives were asserting at the same time that "safety had never been better". This dispute, along with additional contributing factors, led to a significant lack of trust among stakeholders and impeded open and constructive communication on sensitive issues.

Traditionally, the offshore petroleum sector relied on a limited selection of indicators, primarily focusing on the frequency of lost-time incidents to depict safety trends. However, this approach is widely recognized as only providing a partial view of the overall safety picture.

In response to this dispute, Havtil initiated the RNNP project to provide safety information that is impartial and objective as far as practicable, reflecting the current conditions and trends in the Norwegian petroleum sector.

(2) *Objectives*

The primary objective of the RNNP project is to contribute to establishing a realistic and collaboratively agreed picture of trends in HES work, thereby supporting both Havtil and the industry in their efforts to enhance the HES level in petroleum operations. The specific objectives include:

- Measuring the influence of the safety-related work within the industry.
- Contributing to identifying safety–critical areas that require prioritization in identifying causes to prevent undesirable incidents and accidents.

- Enhancing comprehension of the potential causes of undesirable incidents and accidents, along with their relative importance in the context of risk, in order to provide better decision support for both the industry and regulatory authorities pertaining to preventive safety measures and emergency preparedness planning.

(3) *Key limitations*

From the outset, the RNNP project was restricted to what is considered "offshore activity" within Havtil's area of authority. Vessels engaged in offshore operations were excluded, except when operating within a designated 500-m "safety zone." Helicopter operations were similarly restricted to the phases of landing and take-off within this "safety zone". Currently, all passenger transport via helicopter is incorporated into the project, in collaboration with the Civil Aviation Authority of Norway (CAAN) and helicopter operators on the NCS. The project encompasses:

- All production and mobile facilities located on the NCS, encompassing subsea facilities;
- The utilization of vessels within the safety zone surrounding these facilities;
- Helicopter transport of passengers between terminals and facilities.

Onshore facilities have been included in the administrative responsibility of Havtil since 1 January 2004, and data collection started from 1 January 2006. There are eight onshore petroleum facilities, which include two refineries and three large-scale gas processing plants. The rest are relatively small gas terminals and plants. Consequently, experience with onshore facilities is somewhat limited, and most recommendations are derived from offshore installations.

Precursor Indicators

(1) *Precursor data*

The RNNP project has collected precursor data over a span of more than 20 years. Initially, QRA studies and comprehensive reviews were conducted to ensure that no hazards were overlooked. Table 8.1 provides an overview of the categories of defined hazard and accident conditions (referred to as DFUs in Norwegian) associated with major accidents. These DFUs are divided into two primary categories: uncontrolled hydrocarbon releases and fires (DFU1-4), and structure-related incidents (DFU5-10). The values presented encompass all production and mobile drilling units operating on the NCS during the specified timeframe. Table 8.2 illustrates the number of installations and the corresponding man-hours on the NCS, which serve as the foundation for the RNNP project.

The analysis of Table 8.1 indicates a significant annual number of precursors for DFU1, DFU3, and DFU5. Specifically, DFU5 pertains to external merchant vessels that are on a course for potential collision, which implies that such events do not

Table 8.1 Overview of DFUs associated with major accidents

DFU No.	DFU description	Average No./year (2008–2017)	Notes
1	Unignited hydrocarbon leak	11.0	A distinction is made between the non-ignited and ignited hydrocarbon leaks. For instance, unignited gas leaks could cause gas to spread over extensive areas, so that the subsequent ignition could trigger an explosion
2	Ignited hydrocarbon leak	0	
3	Well incidents/loss of well control	16.1	Loss of well control can result in a blowout, which may cause substantial damage to people, the environment, and facilities
4	Fire/explosion in other areas, not hydrocarbon	1.4	A case in point is fires breaking out in living quarters, which could escalate into a major accident
5	Ship on collision course	10.5	A distinction is made between collisions with ships maneuvering close to the facility (supply vessels, shuttle tanks) and collisions with ships not related to the facility, as well as with drifting objects like barges
6	Drifting object	0.7	
7	Collision with field-related vessel/ facility/shuttle tanker	0.2	
8	Damage to a facility's structure, stability/anchoring/ positioning failure	6.9	Platforms and rigs are designed to endure minor collisions. However, significant impacts from a large ship, particularly at high speeds, can result in severe structural damage, potentially leading to total collapse. Additionally, extreme weather conditions can result in such a hazardous situation, including failures in stability and anchoring systems on floating facilities. In the worst case, this can result in the total breakdown of the facility

(continued)

Table 8.1 (continued)

DFU No.	DFU description	Average No./year (2008–2017)	Notes
9	Leak from riser, pipeline and subsea production facility	0.4	A distinction is made between damage to and leakage from risers, pipelines and subsea production facilities Seabed facilities can be damaged by falling objects. Fishing gear can also cause significant damage. The major accident potential primarily stems from oil spills that can cause significant environmental damage. If the subsea facility is located close to a surface installation, it may also be exposed to danger
10	Damage to riser, pipeline and subsea production facility	4.4	

Table 8.2 No. of offshore installations and man-hours in the Norwegian sector in 2021

Installation type	No. of installations, 2021	Man-hours, 2021
Fixed production installations	23	31,637,142
Floating production installations		
• with remote wells	22	
• with wells on or below installation	5	
Production complexes	11	
Normally unattended installations	10	
Mobile installations	19.9	12,499,706
Sum	85.9	44,136,848

accurately reflect the safety management practices of the installation. Notably, the aggregate number of installations exceeds the figures presented in Table 8.2 due to the presence of "complexes" comprising multiple installations. Additionally, the number of mobile installations is determined by averaging the total operational days over 365, as a lot of mobile installations operate in the Norwegian sector for only a portion of the year. A typical offshore working year encompasses 1,460 man-hours, resulting in an approximate total of 20,000 man-years associated with production installations and 10,000 man-years on mobile installations.

(2) *Individual Indicators*

When the number of reported DFUs is collected, it is essential to consider the level of exposure to derive a meaningful representation of trends. If the level of exposure doubles and the incident rate remains constant, the number of DFUs can be expected to double as well. Thus, the normalization of indicators is a crucial process involving the division of incident rates by the volume of exposure to enable standardized comparisons across different situations and contexts.

Since no single measurement can adequately capture all aspects of normalization, it is necessary to utilize an array of parameters, including:

- Man-hours, representing the working hours,
- Number of installation years, categorized by the type of the installation,
- Number of wells drilled, categorized by the type of the drilling activity.

Figure 8.2 illustrates the trend of leaks exceeding 0.1 kg/s, normalized by man-hours for production facilities. These values can also be interpreted as a "relative risk indicator", as they are derived from normalized values rather than absolute values. The final bar in this chart represents a prediction interval for the year 2017, derived from the average level in the preceding period (2008–2016), and reflects a 90% confidence level. These intervals have been computed using the Poisson distribution. If the value for 2017 falls within the lower section of the bar, it signifies a statistically significant reduction in incident rate for the year 2017 compared to the average for the years 2008–2016. An additional example is provided in Fig. 8.3, which depicts the total number of DFUs (1–10) normalized by man-hours.

(3) *Total Risk Indicator*

In the analysis of individual indicators, the typical picture is that some indicators exhibit an upward trend, and others may demonstrate a downward trend. Thus, it is

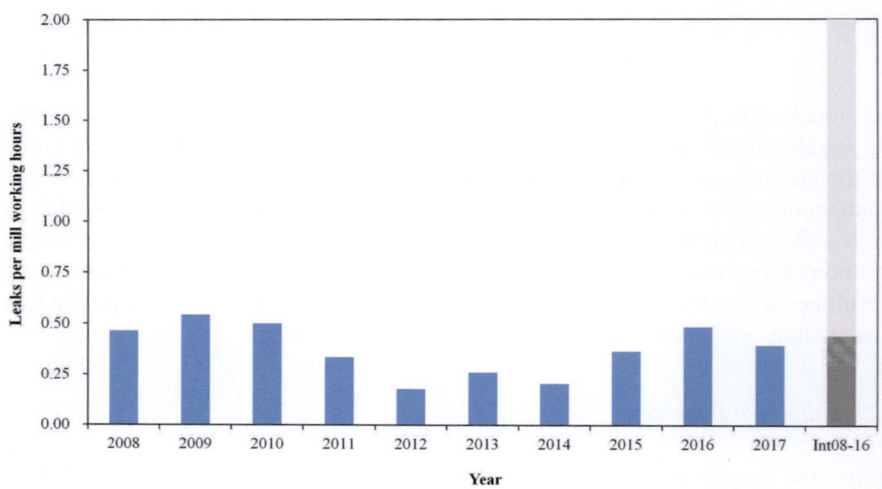

Fig. 8.2 Number of hydrocarbon leaks exceeding 0.1 kg/s, normalized by man-hours

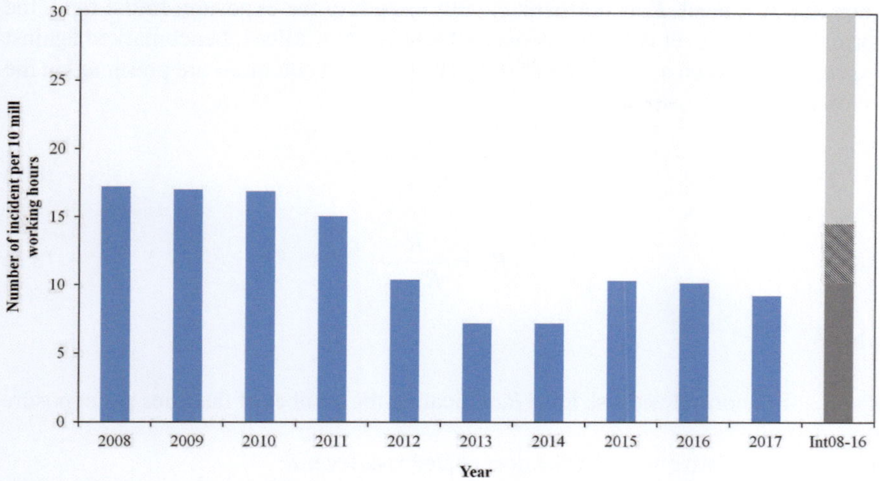

Fig. 8.3 Total number of DFUs (1–10) normalized by man-hours

impossible to obtain a definitive trend. Consequently, it is advantageous to develop a total indicator that can counterbalance the influences of various individual indicators, facilitating a clear understanding of the overall development. The total risk indicator serves as a composite measure, derived from the frequency of incidents and the potential for these incidents to result in fatalities if they escalate into major accidents. This indicates that the total risk indicator assigns weights to each DFU based on their associated potential contributions for loss of life. Notably, the total risk indicator reflects the effects of risk management.

It may be assumed that the total risk indicator would consistently fall within the prediction interval. However, this is not always true, as there are substantial variations in the weights assigned to different DFUs. This suggests that certain DFUs may exert a more pronounced influence than others. The principles for developing the total risk indicator are outlined in the following.

Firstly, the risk level, R, is expressed by

$$R = \sum_i \sum_j \mathrm{DFU}_{ij} v_{ij} = \sum_i \sum_j \mathrm{DFU}_{ij} \mathrm{EX}_{ij} \tag{8.1}$$

where

R	The annual risk level, represented as the expected number of fatalities per year, is based on the number of DFUs associated with the installations;
DFU_{ij}	The annual number of DFUs per DFU type i for installation j;
v_{ij}	The weight assigned to DFU type i at installation j;
EX_{ij}	The statistically expected number of fatalities per incident in DFU type i at installation j.

Then, the risk level, R, is normalized with respect to the exposure. In the final, the normalized risk level is further converted into relative values, benchmarked against a specific year, such as the year of 2005. The pertinent equations are presented in the following:

$$R' = \frac{R}{V} \tag{8.2}$$

$$R'' = \frac{R'}{R'_{2005}} \tag{8.3}$$

where

R'	The normalized risk level R, indicating the number of fatalities per exposure unit;
R''	The relative value of the normalized risk level R;
V	The annual volume of exposure;
R'_{2005}	The normalized risk level R in a specific year, such as the year 2005.

What needs to be emphasized is the way that the weights of DFUs are defined. The definition is associated with the event tree analysis, which is commonly utilized in risk analysis. For a given DFU in an event tree, the end events represent the probability distribution for the possible consequences of the given DFU. By summing up all the end events, the expected value for the given DFU could be obtained. If the consequence is limited to fatalities, the expected value is equivalent to the weight assigned to the given DFU. From a mathematical standpoint, this process involves adding together the products of conditional probabilities for all end events. These conditional probabilities reflect the effectiveness of barriers during a given accident scenario. Thus, the weights are determined by data from QRA studies.

Figure 8.4 illustrates an example of the total risk indicator, depicting the trend for all production and mobile installations, normalized by man-hours. The reference value for the year 2005 is applied for both the total indicator and the three-year rolling average. It is noteworthy that certain years experienced one to two extremely severe precursor events with significant potential for fatalities. These precursor events were assigned individual weights based on their specific contexts, thereby exerting a considerable influence on the value of the total indicator. This discrepancy is a primary reason why the total risk indicator's trend does not closely mirror the trend illustrated in Fig. 8.3. The three-year rolling average serves as a useful tool to mitigate substantial annual fluctuations. The analysis reveals that the underlying trend, when employing a three-year rolling average, indicates a positive trend over time, accompanied by a recent period of stabilization. This trend implies that industry stakeholders have enhanced their management of key factors influencing major accident risk during this period. Furthermore, this underscores the necessity for sharp focus as well as active and continuous management over factors that may affect major accident risk.

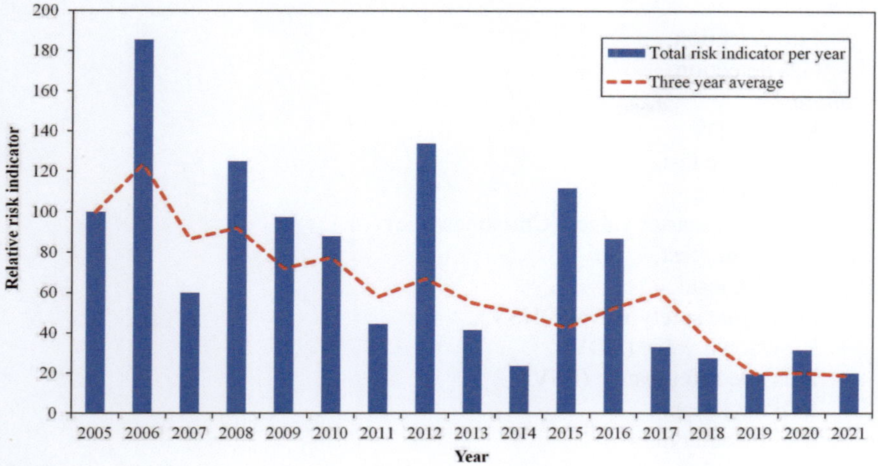

Fig. 8.4 Total indicator for major accidents per year, normalized by man-hours

It is crucial to clarify that the total risk indicator does not serve as a forecast of risk values in the future. Rather, it functions as an indicator that reflects precursor events in the past, which are weighted by their potential fatality risk.

Indicators for Physical Barriers

(1) *Barrier data*

Barriers play a critical role in ensuring safety on the NCS, as clearly outlined in the current management regulations. The purpose of using indicators for barriers is to reflect how the installed barriers respond to potential accidents.

The primary focus is on technical barriers related to leaks within the processing area. Five main barrier functions are encompassed:

- Maintaining the integrity of hydrocarbon production and processing facilities,
- Preventing ignition,
- Reducing clouds and releases,
- Preventing escalation,
- Preventing fatalities.

These barriers comprise various interrelated barrier elements. For example, the detection of a leak is a prerequisite for isolating ignition sources and activating the emergency shutdown valve (ESDV).

A full list of the diverse barrier elements in relation to hydrocarbon production and processing installations, for which the RNNP project collects data, is presented in the following.

- *Detection*:
  - Fire detection,
  - Gas detection.
- *Shutdown*:
  - Riser ESDV
    Closure test,
    Leak test.
  - Wing and master valves (Christmas tree)
    Closure test,
    Leak test.
  - Downhole safety valve (DHSV),
  - Blowdown valve (BDV),
  - Pressure safety valve (PSV),
  - BOP isolation.
- Active fire safety:
  - Deluge valve,
  - Fire pump start.

(2) *Individual barrier indicator*

In the context of the RNNP project, the barrier indicator is defined by the failure fraction derived from the test data in association with the performance of barriers against major accidents on the installations. Failure fraction can be expressed through two different approaches: overall percentage of failures (OPF) and average percentage of failures (APF).

The OPF represents the ratio of the total number of all failures for all reporting installations to the total number of all tests conducted at these installations. This approach ensures an equal contribution of each test to the average. In this way, installations that have performed many tests will largely dominate the results. Thus, the OPF will primarily reflect the performance of barriers at installations with many tests, rather than providing a representative assessment for the NCS.

The APF is defined by the percentage of failures over tests for each installation individually, and then is averaged for all installations. This approach ensures that each installation contributes equally to the overall average, irrespective of the number of tests conducted at any single installation. This prevents installations that perform many tests from dominating the results. However, it introduces the challenge of statistically unreliable data from the installations that have conducted only a limited number of tests.

The pertinent mathematical formulations are outlined in the following.

$$FF_{\text{overall}} = \frac{\sum_{j=1}^{n} x_j}{\sum_{j=1}^{n} y_j} \tag{8.4}$$

$$FF_{\text{average}} = \frac{1}{n} \sum_{j=1}^{n} \frac{x_j}{y_j} \tag{8.5}$$

where the following notations are employed:

FF The failure fraction over tests
x_j The number of failures associated with the installation j
y_j The number of tests associated with the installation j
n The number of installations that have performed tests for the barriers.

Figure 8.5 shows the overall and average percentages of failures for barrier elements in the year 2019. It indicates that for three barriers out of ten, the OPF is higher than the APF. Furthermore, it is important to note that an industry norm has been established for most barrier elements, allowing for a comparative analysis of the barrier indicators against these norms. The findings indicate that the APF for the riser ESDV, DHSV, BDV, deluge valve and fire pump start all exceed the industry norm for availability. Notably, there is a lack of an industry norm for the use of BOP in isolation procedures. In such a case, it is recommended to conduct a trend analysis to monitor failures associated with this BOP barrier.

(3) *Total barrier indicator*

When presenting trends for different barrier elements, it frequently becomes challenging to discern an overall trend. It has been found that certain barriers may exhibit a positive trend, while others may demonstrate negative trends or lack a clear trend. Analogous to the total risk indicator for potential major accidents, it is also beneficial to possess a total barrier indicator that counterbalances the influences of individual barrier indicators and reflects the overall development of the barrier performance.

The total barrier indicator in the RNNP project is developed on the basis of the frequency of incidents pertaining to loss of containment, with simultaneous failures of the following barrier functions:

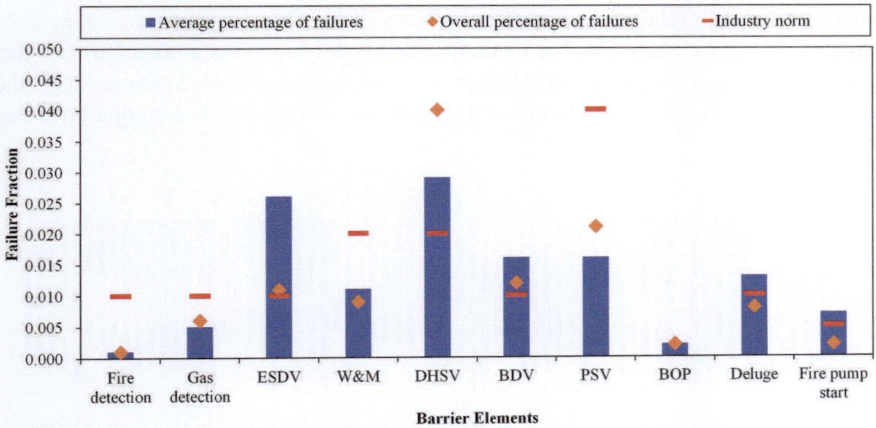

Fig. 8.5 The overall and average percentages of failures for barrier elements in 2019

- Detection through gas detection and fire detection systems.
- Isolation and content minimization of segments according to riser ESDVs, well isolation using Christmas tree valves, DHSV, BOP, BDVs and PSVs.
- Activation of the fire water system through initiating the fire water supply pumps and opening the deluge valves.

The pertinent mathematical formulations for calculating the total barrier indicator are presented in the following:

$$\mathrm{TBI}_i = f_{\mathrm{leak},i} \cdot \prod_{j=1}^{n} \left(\frac{x_{ij}}{y_{ij}} \cdot w_{ij} \right) \tag{8.6}$$

where the following notations are used:

TBI_i	The total barrier indicator for the installation i
$f_{\mathrm{leak},i}$	The frequency of hydrocarbon leaks exceeding 0.1 kg/s on the installation i
x_{ij}	The number of failures in tests of the barrier function j on the installation i
y_{ij}	The number of tests for the barrier function j on the installation i
w_{ij}	The weight of the barrier function j on the installation i
n	The total number of barrier functions on the installation i.

Figure 8.6 illustrates an example of the total barrier indicator, intended to represent the overall performance of all individual barriers. The values are displayed on a logarithmic scale and the exact numerical values are not shown. Each bar is color-coded to signify different barrier functions, arranged sequentially from top to bottom as follows: containment, fire detection, gas detection, isolation, firefighting, and mustering. The depth of each bar indicates the performance level of the corresponding barrier. Subsequently, the contributions from all individual barrier functions are aggregated to provide a comprehensive assessment of overall barrier performance in consideration of all barrier elements. For example, the barrier functions associated with the installation G exhibit greater depth than those of the installation F, suggesting that the installation G demonstrates superior overall barrier performance compared to the installation F.

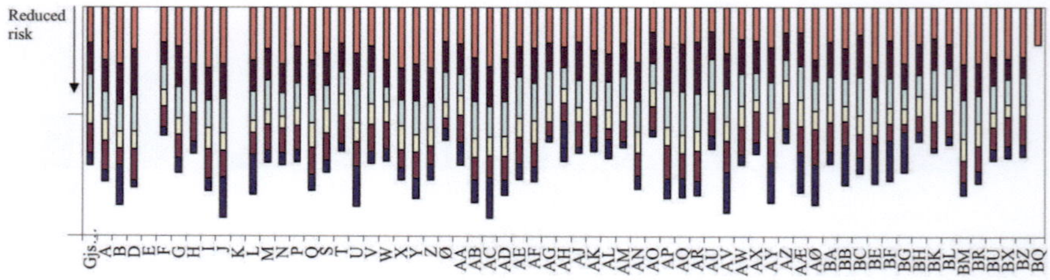

Fig. 8.6 The total barrier indicator for anonymous installations

When interpreting the results of the total barrier indicator for each installation, the following limitations should be noted.

- Only a select few barrier elements, which are easily testable, are incorporated.
- The actual configuration of the barrier system, including the redundancies and capacities, is not incorporated.
- Only the barrier performance related to availability on demand is incorporated, whereas effectiveness, capacity and robustness are excluded.

Organizational Indicators

The RNNP project consists of a comprehensive questionnaire-based survey conducted every two years, focusing on the HES climate, working environment conditions, as well as the perceived risk of employees.

In 2021, a total of 6378 individuals participated in the survey, representing 25.9% of the estimated workforce. The overall findings indicate a negative development in the HES climate, working environment factors, and health complaints from 2019 to 2021.

The RNNP survey can be used as an overall indicator of important HES climate parameters, such as the statement "I find it unpleasant to point out that safety rules and procedures have been broken". For well drilling crews on all installations on the NCS, the average score in the last four questionnaire surveys is as follows:

- 2015:3.8,[1]
- 2017:3.7,
- 2019:3.7,
- 2021:3.6.

This shows that the pressure to accept deviations from safety rules and procedures within drilling operations has been increasing in the past years, which may be considered a clear sign of increased unsafe behaviour on the installations.

8.4.2 Major Accident Risk Indicators at the Enterprise Level

The preceding section has delineated major accident risk indicators at the national level, drawing upon data and experience from the RNNP project initiated by Havtil. A primary objective of this initiative is to encourage individual enterprises to formulate their own schemes for major accident risk indicators, thus preventing them from facing the same situations that preceded numerous offshore disasters, where monitoring was restricted to HES indicators pertaining to occupational

[1] The value 3 corresponds to "indifferent" and the value 4 corresponds to "disagree", thus the trend is in the direction of agreement.

accidents. The RNNP project serves as a valuable starting point for formulating such schemes. This section elaborates on how enterprises can develop their own major accident risk indicator schemes, drawing on the experiences from large-scale offshore oil enterprises, such as Equinor and Total Energies Norway.

Precursor Indicators

At the enterprise level, precursor indicators should not be entirely overlooked, as they offer valuable insights into organizational performance over time and reflect the outcomes of undesirable events.

The biggest challenge with the development of precursor indicators arises in small and medium-sized enterprises with limited installations, as the infrequency of precursor incidents renders the establishment of reliable indicators for major accident risk trends. For example, only five precursor incidents were recorded in the RNNP project for a small-sized enterprise operating on the NCS over a 5-year period. From the perspective of the authors, the precursor data sources could be extended from other national sectors of the small and medium-sized enterprises in the current situation. Provided the total number of installations reaches a sufficiently high level, this approach could yield viable solutions.

In contrast, the large-scale oil enterprises typically possess an adequate number of installations, allowing for the generation of meaningful precursor indicators at the enterprise level.

However, there exists an exception. The precursor indicator regarding the frequency of hydrocarbon leaks with flow rates above 0.1 kg/s (DFU 1), warrants particular attention, though its occurrence frequency is low. As illustrated in Fig. 8.7, an average installation will experience a reportable leak once every few years, a frequency insufficient to establish meaningful trends. Nevertheless, this indicator should be taken seriously due to the great concern within the industry. Notably, certain installations may report several leaks annually, exceeding the average incident rate by over ten times. Such a case indicates the importance of this indicator under certain circumstances.

For individual enterprises, the following precursor indicators are recommended.

- As a minimum, indicators based on the occurrence of hydrocarbon leaks for each installation.
- As a minimum, indicators based on the occurrence of DFUs from the RNNP project if the data sources are sufficient.

Indicators for Physical Barriers

In practice, leading indicators for major accident risk are derived from performance data pertaining to barrier elements. Theoretically, these barrier elements can be both physical (technical) and non-physical (organizational and operational). However, to this day, the enterprises have been primarily focusing on indicators related to technical barrier elements.

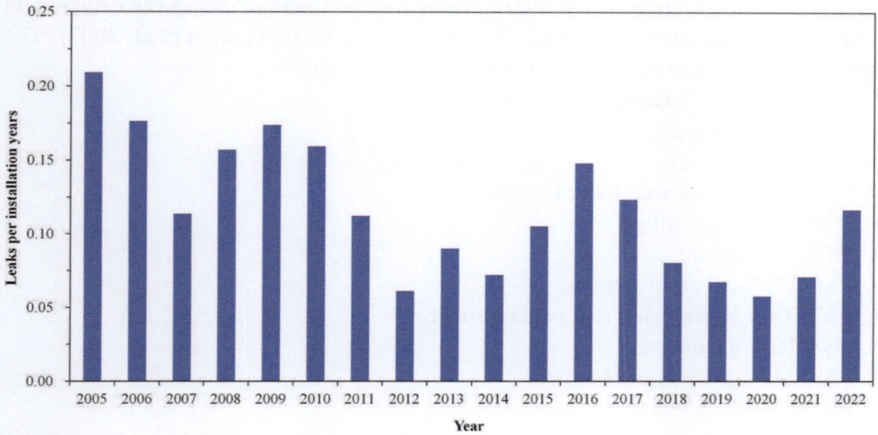

Fig. 8.7 Number of leaks, normalized against installation years, all installations

Typically, there exists a sufficient volume of data concerning technical barriers, enabling the follow-up of individual installations. This suggests that technical barrier indicators are meaningful at both the installation and enterprise levels.

What calls for special attention is that the reporting of safety-critical failures associated with barriers is conducted with a high-quality level to prevent both under-reporting and over-reporting. Practice has proven that the problem of over-reporting is significantly more serious than that of under-reporting. For example, if an ESDV fails to close on demand, this failure is classified as safety-critical. In contrast, provided an ESDV can to be re-opened, it could pose a risk to production assurance, but it does not qualify as a safety-critical failure and should therefore be excluded from the reporting scheme of barrier indicators.

Barrier indicators could be followed up independently for each installation and the enterprise overall. The primary use of these indicators is to facilitate tracking trends over long periods. In Norway, certain enterprises have collaborated to develop target values that serve as benchmarks for the observed values. These target values are crucial for defining and assessing the status of barriers.

For individual enterprises, the following technical barrier indicators for each installation are recommended:

- The minimum requirement is to incorporate technical barrier elements, which are adopted by the RNNP project.
- There could be additional technical barrier elements, if applicable.

Indicators for Non-physical Barriers

There has been continued interest in developing indicators for non-physical barriers. However, few successful stories are known in this area. The authors have proposed and defined a series of performance standards (PS) and performance requirements

(PR), which are grounded in the International Atomic Energy Agency (IAEA) project (IAEA 2000) and the OTS project (Kongsvik et al. 2010; Vinnem et al. 2007). The defined performance standards include (Zhen et al. 2019):

- PS1: Work practice,
- PS2: Competence,
- PS3: Procedures and documentation,
- PS4: Communication,
- PS5: Workload,
- PS6: Physical environment,
- PS7: Task supervision and management,
- PS8: Risk awareness,
- PS9: Management of change,
- PS10: Work scheduling.

Examples of performance requirements for PS2 (Competence) are presented in Table 8.3. The authors have developed the quantification rules (Zhen et al. 2019) for measuring the status of performance standards and performance requirements. However, at present, neither the performance standard nor the performance requirement can be considered appropriate as indicators. This limitation primarily arises from the absence of straightforward and effective methods for measuring the levels of such indicators.

So far, there are still no well-established indicators for non-physical barriers. Nevertheless, efforts are ongoing to identify parameters that may serve as suitable

Table 8.3 Examples of performance requirements for "PS2: competence"

PR type	No.	Definition and delineation
General requirements	2.1	The management and the operational team must possess comprehensive knowledge and capabilities in the following areas: (1) major accident risks, (2) safety-critical activities, (3) processes and operations, and (4) management of deviation scenarios
Specific requirements	2.1.1	The teams possess the necessary expertise and capabilities to carry out critical tasks safely
	2.1.2	The contractors receive adequate training to perform their responsibilities safely
	2.1.3	The training provided is sufficient to address deviations and emergency situations effectively
	2.1.4	The expertise level in understanding processes and system is adequate for the tasks being undertaken
	2.1.5	The management possesses the necessary capability to effectively monitor and oversee all safety-critical activities

indicators for non-physical barriers. Though this work is highly challenging, it is widely recognized that non-physical (organizational and operational) barriers require equal or even greater emphasis compared to technical barriers. Thorough investigations into major accidents have confirmed that most of the deviations and underlying causes are attributable to human and organizational factors.

8.5 Questions

Q1: Why should leading indicators be preferred over lagging indicators?
Q2: Why are hydrocarbon leaks below 0.1 kg/s not considered a good indicator for major accident risk?
Q3: What is the advantage of major accident risk indicators at a national level?
Q4: What is the advantage of major accident risk indicators at an enterprise level?
Q5: Why can not precursor indicators be used to assess the performance of barriers?

References

Dyreborg J (2009) The causal relation between lead and lag indicators. Saf Sci 47(4):474–475
Erikson S (2009) Performance indicators. Saf Sci 47(4):468
Hale A (2009) Editorial special issue on process safety indicators. Saf Sci 47(4):459
Haugen S, Seljelid J, Mo K, Nyheim OM (2011) Major accident indicators for monitoring and predicting risk levels. In: SPE european health, safety and environmental conference in oil and gas exploration and production. Vienna, Austria
Havtil (2024) Trends in risk level in the petroleum industry on the Norwegian Continental Shelf. Main Report. Stavanger, Norway
Heinrich HW (1931) Industrial accident prevention. McGraw-Hill, New York
Hopkins A (2009) Thinking about process safety indicators. Saf Sci 47(4):460–465
HSE (2006) Developing process safety indicators: a step-by-step guide for chemical and major hazard industries. Surrey, UK
Hudson PTW (2009) Process indicators: managing safety by the numbers. Saf Sci 47(4):483–485
International Atomic Energy Agency (IAEA) (2000) Operational safety performance indicators for nuclear power plants. Report No. IAEA-TECDOC-1141, Vienna, Austria
Kjellén U (2000) Prevention of accidents through experience feedback. CRC Press
Kjellén U (2009) The safety measurement problem revisited. Saf Sci 47(4):486–489
Kongsvik T, Almklov P, Fenstad J (2010) Organizational safety indicators: some conceptual considerations and a supplementary qualitative approach. Saf Sci 48:1402–1411
Øien K, Utne I, Herrera I (2011) Building safety indicators: part I-theoretical foundations. Saf Sci 49:148–161
Pasman H (2015) Risk analysis and control for industrial processes gas oil and chemicals. IChemE Botterworth Heineman, Oxford
Pasman H, Rogers W (2014) How can we use the information provided by process safety performance indicators? Possibilities and limitations. J Loss Prev Process Ind 30:197–206
Visser JE (1995) Managing safety in the oil industry-the way ahead. In: Proceedings of loss prevention and safety promotion in the process industries. Antwerp, Belgium

Vinnem JE (2000) Risk monitoring for major hazards. In: SPE international conference on health, safety and environment in oil and gas exploration and production. Stavanger, Norway

Vinnem JE (2010) Risk indicators for major hazards on offshore installations. Saf Sci 48(6):770–787

Vinnem JE, Aven T, Sørum M, Øien K (2003) Structured approach to risk indicators for major hazards. In: European safety and reliability conference. Maastricht, The Netherlands

Vinnem JE, Aven T, Husebø T, Seljelid J, Tveit OJ (2006) Major hazard risk indicators for monitoring of trends in the Norwegian offshore petroleum sector. Reliab Eng Syst Saf 91(7):778–791

Vinnem JE, Haugen S, Kongsvik T, Seljelid J, Steen S, Sklet S, Thomassen O (2007) Operational safety condition-concept development. In: European safety and reliability conference. Stavanger, Norway

Zhen XW, Vinnem JE, Næss S (2019) Building safety in the offshore petroleum industry: development of risk-based major hazard risk indicators at a national level. Process Saf Environ Prot 128:295–306

Zhen XW, Vinnem JE, Han Y, Peng CY, Huang Y (2022) Developments and prospects of major accident indicators in the offshore petroleum sector. Process Saf Environ Prot 160:551–562

Ninth Lesson: Risk-Informed Decision-Making Framework

9

Nothing is more difficult, and therefore more precious, than to be able to decide.

—Napoleon Bonaparte (1899)

Abstract

The risk management has generated considerable expectations from enterprises as it could effectively balance the benefits of opportunities with accident prevention. This lesson focuses on the decision-making framework for risk management. Firstly, the need for risk-informed decision-making is discussed, exploring why the conventional mechanized approach falls short in the decision-making process. Then, the challenges of current decision-making frameworks are presented. On this basis, the fundamental components of the decision-making framework are discussed. These components underpin the justification and appropriateness of the decision-making framework. The central focus of the lesson delves into a suggested decision-making framework. In the final, the lesson explores how the decision-making framework is influenced by its framework conditions. This lesson provides a theoretical basis and practical guidance for readers on how to make optimal decisions.

9.1 Needs and Challenges for Risk-Informed Decision-Making

There has been a debate regarding how to conduct decision-making within the framework of health, environment, and safety (HES) management. One viewpoint is that the decision-making process should be highly mechanized through the use

X. Zhen and J.-E. Vinnem, *Ten Lessons on Risks*,
https://doi.org/10.1007/978-3-031-99064-9_9

of Risk Tolerance Criteria (RTC). This mechanistic approach suggests that the decision-making process will not be dependent on the involved personnel and their abilities. Thus, it allows for a rapid reaction to known situations and results in better predictability as fewer subjective judgments are involved in the decision-making. It is generally believed that cost minimization could be achieved at least in the short term. Nonetheless, this mechanistic approach would fail to recognize the important role of management in performing the complex value judgments involving uncertainties.

The authors, along with many other scholars, hold the view that the HES decision-making process should adopt a comprehensive framework that incorporates all pertinent factors as inputs to decision-making. Decisions should be ultimately made by management following a careful evaluation and judgment of available options. Further, it is widely recognized that risk cannot be entirely eliminated and must be effectively managed. Thus, the comprehensive framework, which is recommended and given great expectations for decision-making, is grounded in risk management. The risk-informed decision-making should be able to obtain a certain level of consistency in decision-making and confidence in obtaining desirable outcomes.

In addition, it should be noted that the current decision-making frameworks encounter multiple challenges, as pointed out by Aven et al. (2007). These challenges include:

- The fundamental comprehension of the decision-making framework, involving the use of statistical expected values and the treatment of uncertainties;
- The use of diverse tools for decision and risk analysis, including sensitivity analysis and robustness analysis;
- A structured approach to categorizing the decision context to simplify the decision-making process;
- The way of conducting trade-offs to prioritize objectives and balance various factors, such as costs, risk, etc.;
- The way of reflecting interactions and dependencies among decision problems at different project stages and authority levels;
- The overall communication and presentation of the decision basis.

In response to these challenges, a decision-making framework grounded in risk management is presented and discussed in this lesson, drawing on the works of Aven and Vinnem (2005), Aven et al. (2007), Oil and Gas UK (2014) and Vinnem and Aven (2006).

9.2 Components of the Risk-Informed Decision-Making Framework

This section delineates and examines the fundamental components of a risk-informed decision-making framework. These components will serve as the premises that underpin the justification and suitability of the framework.

(1) HAZID

A meticulous and comprehensive identification of hazards is paramount within the risk management process. The inability to recognize a hazard may fundamentally flaw the final decision.

(2) Risk

A comprehensive viewpoint on risk is adopted, acknowledging that risk cannot be distinguished from the decision context. Risk is a part of the issues being addressed, the people who assess the risk, the methods and tools employed, and so on. Attention should be given to risk transfer and the limited information on the risk representation, such as the use of generic data.

(3) Risk tolerability

The risk level should, as a minimum, remain below the upper tolerability limit and be As Low As Reasonably Practicable (ALARP), which is essential for the application of the decision-making framework. Thus, the decision-making framework is consistent with Norwegian and UK regulations, both of which have references to risk reduction in line with ALARP. Notably, the risk tolerability cannot merely be determined based on the risk analysis results. Rather than accepting the risk itself, we accept a solution with all its attributes, such as cost, profit, loss of life, and time.

(4) Uncertainty

A critical component of the decision-making framework is that any risk-informed decision must be made with an adequate level of certainty. This is particularly important in cases where a safer alternative is not selected, or there is significant uncertainty allowing for a possibility that the selected alternative may not meet the RTC. In evaluating risk tolerability, it is crucial to demonstrate that the margin between the evaluated risk level and the RTC exceeds the level of any uncertainty. In scenarios where severe consequences are possible, the risk assessment should be conservative to ensure that the uncertainties will not result in a less safe alternative being selected. It is essential to clarify that a low level of uncertainty does not necessarily indicate a low risk level, and correspondingly, a high level of uncertainty does not automatically indicate a high-risk level. The NORSOK Z-013 standard (Standards Norway 2023) outlines a methodology for the critical assessment of assumptions.

(5) Priority of risk-reducing measures

Although the ultimate outcomes of risk reduction may be the same, hazard avoidance and accident prevention are prioritized over control and mitigation of accident consequences. The recommended priority of risk-reducing measures is as follows:

- Elimination or minimization of hazards through design, concentrating on inherently safe design;
- Prevention of accidents, concentrating on measures that reduce the probability of accidents occurring;
- Mitigation of consequences, concentrating on measures that minimize the severity of accidents;
- Establishment of evacuation, escape and rescue (EER) arrangements.

It can be noted that the fundamental components of the risk-informed decision-making framework are closely linked to the manageability of risk, the concept of which is illustrated in Fig. 9.1. It is evident that certain risks, such as the Alternative B, are more manageable compared to other risks, such as the Alternative A. That is to say, in comparison to the Alternative A, although Alternative B is initially associated with a higher risk, it presents a greater potential for reducing the risk level. Figure 9.2 illustrates how the degree of uncertainty and manageability evolve over time, starting from the project's inception. The focus of real observations, denoted as $X = (X_1, X_2, \ldots)$, can be predicted using the expected value $E[X|K]$, where K represents the background knowledge. As shown in Fig. 9.2, the degree of uncertainty and manageability is dependent upon the development stage of a system. The degree of uncertainty and manageability tends to be higher in the early stages and decreases as time progresses. Typically, decisions are made during the early stages when it is facing significant uncertainties of possible consequences associated with an activity. Then, this early decision imposes constraints on the further development of the system, and the degree of uncertainty and manageability will normally be reduced. When the process progresses, new decisions are made, introducing additional constraints for the system's further development. It is important to note that Fig. 9.2 merely illustrates the typical trends of the degree of uncertainty and manageability over time. In practice, the actual observations X may deviate significantly from these predictions due to significant uncertainties. Nonetheless, the desirable outcomes can still be achieved in accordance with appropriate uncertainty management.

9.3 Suggested Framework for Risk-Informed Decision-Making

The suggested framework for risk-informed decision-making is illustrated in Fig. 9.3, consisting of four primary steps. The first step in implementing this framework is to establish the decision context. Then, the starting point of a decision process is that the decision-maker is facing some decision problem, such as

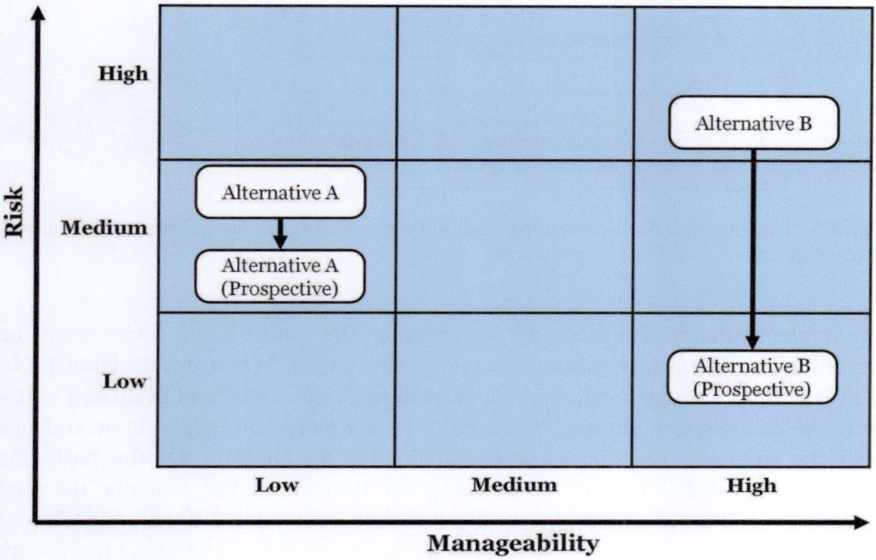

Fig. 9.1 The concept of the manageability of the risk. Adapted from Aven et al. (2007) with Elsevier permission

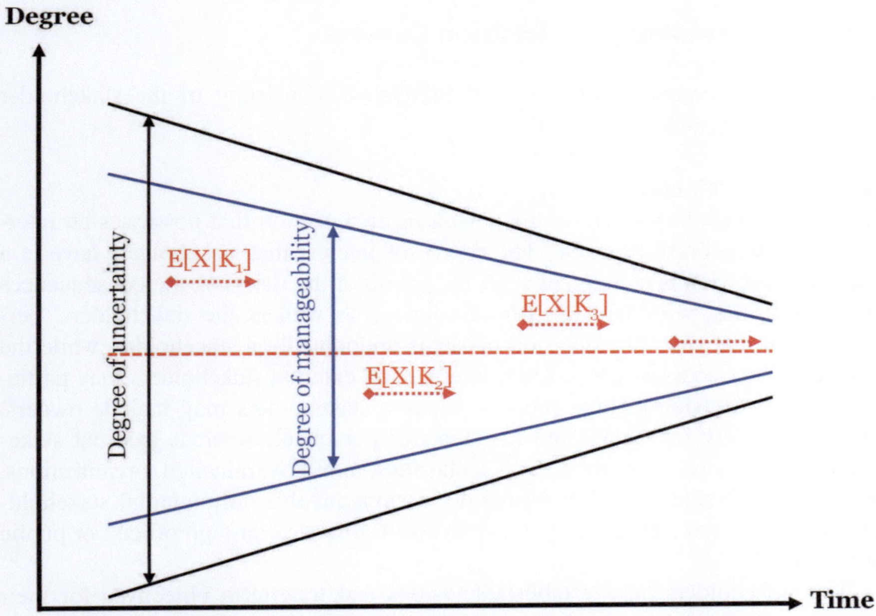

Fig. 9.2 The illustration of the degree of uncertainty and manageability over time. Adapted from Aven et al. (2007) with Elsevier permission

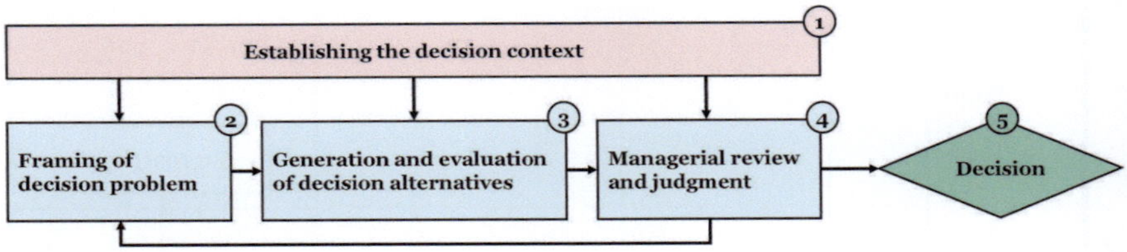

Fig. 9.3 The suggested decision-making framework grounded in risk management. Adapted from Aven et al. (2007) with Elsevier permission

the implementation of a risk-reducing measure, the use of a new technology, the modification of existing process facilities, and so on. When the decision problem is defined, the decision alternatives need to be generated and evaluated. After that, the decision-support analyses need to be reviewed and judged in accordance with the premises, assumptions and limitations of the analyses. On this basis, the risk-informed decision could be made by the decision-maker. Notably, the final decision will depend on how the decision-maker weighs the different concerns.

The rest of this section will delineate the suggested decision-making framework in a stepwise manner, with particular emphasis on the third step, which involves the generation and evaluation of decision alternatives.

9.3.1 Establishing the Decision Context

The decision context should first be established, consisting of the stakeholder influence and decision principles.

Stakeholder Influence

Here, the stakeholder refers to the person or organization that possesses an interest in the decision to be made. The degree of interest that stakeholders have in a particular decision is contingent upon the nature of the risk (i.e., the consequences of the activities, with associated uncertainties), as well as the stakeholders' perceptions of that risk. The decision-maker is undoubtedly a stakeholder, while the reverse is not necessarily true. Both internal and external stakeholders may participate in the decision-making process. Internal stakeholders may include owners, operators, safety managers, and maintenance personnel, whereas external stakeholders may consist of regulatory authorities, non-governmental organizations, and research institutions. The formal discussions involve only internal stakeholders, while external stakeholders primarily exert influence through official or public supervision.

The stakeholders have established visions and long-term objectives for their activities, supported by strategies and plans formulated to achieve them, thereby reflecting their underlying values.

Decision Principles
On the basis of the stakeholder influence, the decision principles can be developed to guide decisions toward the intended outcomes.

Examples of these decision principles include:

- Evaluation of enterprise attributes, which may encompass factors such as investments, profitability, occupational health, environmental sustainability, social responsibility, and reputation;
- High-level decision-making principles, which may involve consensus-building, inclusion of third-party perspectives, and adherence to precautionary and cautionary principles;
- Overall procedures for conducting the decision-making process and implementing the ALARP principle.

9.3.2 Framing of the Decision Problem

This step concentrates on the definition of goals and the decision problem.

Goals Definition
High-level goals pertaining to the project activities and the decisions within the enterprise are formulated. For instance, a typical goal for an enterprise regarding risk tolerance is the implementation of the ALARP principle, ensuring that the risk levels associated with the personnel, environment, and assets are all reduced to ALARP.

Problem Definition

At this stage, the definition of the decision problem encompasses the following aspects:

- The type of the decision problem.
- The specific decision problem to be solved.
- The results to be obtained by solving the decision problem.
- The pertinent criteria and prerequisites for solving the decision problem.

In this book, three types of decision problems are introduced as follows.

(1) Standard Problem
  - The problem is characterized by limited expected loss and uncertainties.
  - Detailed quantitative analysis is not required.
  - External expertise is not required for facilitating and documenting the decision process.

(2) Advanced Problem
  - The problem is characterized by significant expected loss and uncertainties.
  - Detailed quantitative analysis on certain issues is required.

- External expertise is required for facilitating and documenting the decision process.

(3) Complex Problem
- The problem is characterized by substantial expected loss and uncertainties.
- Detailed quantitative analysis is required.
- Competent personnel should be engaged in the management team for facilitating and documenting the decision process. Here, the competence includes but is not limited to good knowledge of the installation and its operating procedures, as well as the QRA techniques.

A classification matrix is developed to ascertain the type of decision problem, as illustrated in Fig. 9.4. The classification rule is developed on the basis of two dimensions, i.e., the expected consequences and uncertainties, as detailed in Tables 9.1 and 9.2. It can be seen that each dimension is associated with a specific set of features. Notably, the total ranking of either the expected consequence or uncertainty will be determined by the highest ranking of any relevant feature.

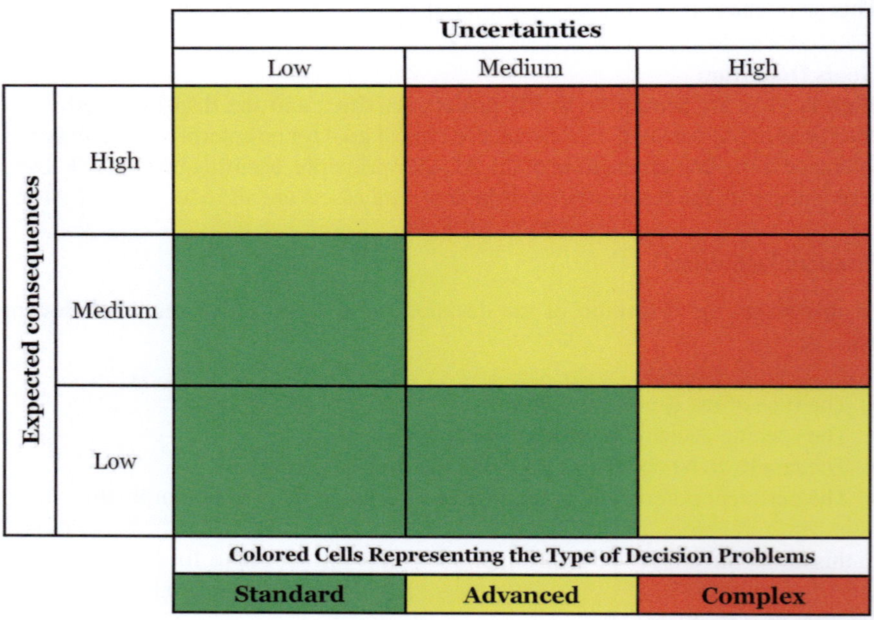

Fig. 9.4 Classification matrix for determining the type of the decision problem. Adapted from Aven et al. (2007) with Elsevier permission

Table 9.1 Ranking criterion of expected consequences, adapted from Aven et al. (2007)

Feature	Ranking (NOK)		
	Low	Medium	High
Investment	>50 mill	1000 mill > $x \geq$ 50 mill	$x \geq$ 1000 mill
Production increase	>50 mill	1000 mill > $x \geq$ 50 mill	$x \geq$ 1000 mill
Production loss	>50 mill	1000 mill > $x \geq$ 50 mill	$x \geq$ 1000 mill
Project schedule	Minor	Serious	Major
Loss of lives	Minor	Serious	Major
Environment	Minor	Serious	Major
Reputation	Minor	Serious	Major
Total	The total ranking is determined by the highest ranking in one of the features		

Table 9.2 Ranking criterion of uncertainties, adapted from Aven et al. (2007)

Feature	Ranking		
	Low	Medium	High
Complexity of technology	Limited	Significant	Large
Complexity of organization	Limited	Significant	Large
Availability of information	Large	Significant	Limited
Time frame	Limited	Significant	Large
Vulnerabilities	Limited	Significant	Large
Total	The total ranking is determined by the highest ranking in one of the features		

9.3.3 Generation and Evaluation of Decision Alternatives

Generation of Decision Alternatives

In addressing a specified decision problem, a management-assigned task force group may propose a range of decision alternatives.

The procedure pertaining to the generation of decision alternatives can vary in complexity. In some cases, the possible alternatives are readily apparent. For example, the decision problem is whether to implement a risk-reducing measure or not. In other circumstances, there could be many possible solutions to the defined decision problem, and each solution may also have a subset of alternatives.

Decision-Making Approaches

Once the decision alternatives have been generated, it is essential for the task force group to employ decision-making approaches to further analyze and evaluate the decision alternatives. The Oil and Gas UK (2014) outlines three primary decision approaches:

(1) Good Practice

According to the Commission for Energy Regulation (2013), good practice refers to the risk management measures that are reasonably practicable for situations where hazards are well understood. Good practice is often codified in established standards and guidelines.

In the situations where established good practice exists, good practice needs to be followed in all circumstances. Thus, if a list of risk-reducing measures is proposed, only the measures that meet or exceed relevant good practice can be taken forward.

Note that good practice is not fixed and it may change over time. This change is primarily driven by the increased knowledge of the hazards, and the change in the RTC. In addition, risk assessment may be incorporated in some good practice.

(2) Risk Assessment

In the situations where good practice is not well-defined or there is no relevant good practice, risk assessment that considers the specific circumstances of the decision is required. The main techniques for risk assessment can be referred to the fourth lesson of this book. It is worth noting that if different results arise from different techniques, the results cannot be assigned weights to arrive at a final decision. This case indicates that either the assumptions and data used should be re-examined, or the use of the techniques themselves should be reconsidered.

The risk assessment must be carried out in a way that decisions can be made with sufficient confidence, of which risks have been reduced to a level that is ALARP. It can be challenging to determine whether the risk level is ALARP or not, as it requires professional judgment from the task force group.

In some cases, a formal cost-benefit analysis (CBA) is further needed to compare the benefits of risk reduction against the associated costs.

(3) Precautionary Approach

In the situations where substantial uncertainties still exist in the risk assessment, which has considered all the available scientific and engineering evidence by competent personnel, a precautionary approach to hazard management should be adopted. A precautionary approach indicates that uncertainty analysis is replaced by conservative assumptions, which will intrinsically ensure the treatment of uncertainty is biased towards safety.

According to the UK Health and Safety Executive (HSE 2001), the precautionary principle ought to be applied in situations where:

- There is a good reason, based on empirical evidence or a plausible causal hypothesis, to believe that serious harm might occur, even if the likelihood of harm is remote;

- The scientific information gathered at this stage regarding consequences and likelihood reveals a level of uncertainty so high that it is impossible to evaluate the conjectured outcomes with sufficient confidence to proceed to the next stages of the risk assessment process.

The precautionary principle addresses hazards associated with high scientific uncertainty and negates the absence of scientific certainty as a justification for not taking preventive actions (HSE 2001). There are two important issues to note in the precautionary principle. For one thing, the assessed hazards should cover the worst-case scenario, but should not include the hypothetical hazards without any credible evidence. For another, safety is expected to receive precedence over economic considerations. This indicates that the risk-reducing measures may be implemented even when the cost is grossly disproportionate to the benefits of the risk reduction, particularly when the risk level associated with non-implementation cannot able to be proved to be ALARP with sufficient certainty.

Evaluation of Decision Alternatives

The approach employed for evaluating decision alternatives is contingent upon the type of the decision problem, as illustrated in Fig. 9.5, which indicates that:

- For a standard decision problem, the decision alternatives will generally be evaluated by the established good practice. If such good practice is not well-defined or there is no relevant good practice, risk assessment is required to evaluate decision alternatives.
- For anadvanced decision problem, there will be a need for risk assessment to evaluate the decision alternatives and ensure that risks have been reduced to a level ALARP. In this case, relevant good practice will still have to be followed.
- For a complex decision problem, the precautionary approach to hazard management is likely to be needed to evaluate the decision alternatives. In this case, relevant good practice must still be followed and risk assessment is needed to inform this approach.

What calls for special attention is that in practice, there is a continuum of decision problems ranging from the standard to the complex. Thus, the evaluation of decision alternatives cannot rely on one of the aforementioned decision approaches merely. Under any circumstances, relevant good practice must be followed and the results of risk assessment must not exceed the upper tolerability limit.

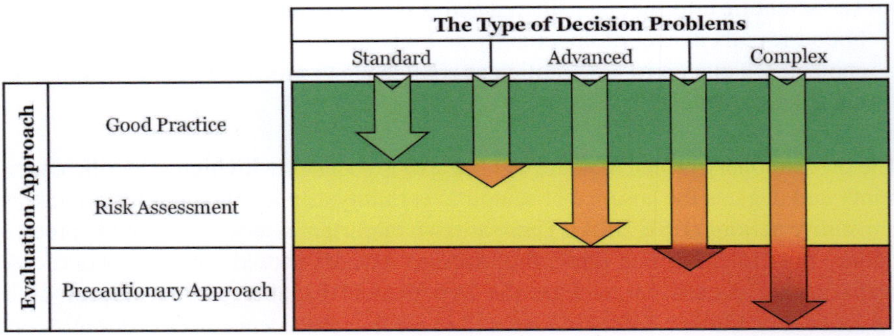

Fig. 9.5 The suggested decision approaches for the corresponding decision problems. *Source* Adapted from ISO 17776 (2016)

9.3.4 Managerial Review and Judgment

A purely mechanical procedure of interpreting the results of decision analyses and evaluations into a final decision cannot be justified. The decision analyses and evaluations provide recommendations on which decision alternative to choose. That is to say, the decision analyses and evaluations provide decision support rather than a decision. The management will consider the broader decision support, including both the quantitative and qualitative input from the task force group. Before making the final judgment on the most appropriate alternative, the following issues need to be reviewed carefully:

- Does the decision-making process follow the established decision principles?
- What is the order of precedence for the decision alternatives based on the decision analyses and evaluations? What criteria are used for ranking?
- What are the premises, assumptions, and limitations associated with the decision analyses and evaluations?
- Are there any pertinent issues that have not been considered in the decision analyses and evaluations?
- Are there any relevant stakeholders not taken into account? How much influence do different stakeholders have on the conclusion?
- How robust is the decision that is to be made?

9.4 Importance of the Decision-Making Framework Conditions

The approach outlined in the preceding sections was developed during a period when the framework conditions within the Norwegian offshore industry were quite different from those observed in the mid-2020s.

At the beginning of the century, there were mainly large international oil enterprises operating on the NCS, with the two Norwegian oil enterprises Equinor and Norsk Hydro challenging each other.

The high-risk levels in the 1970s and 1980s were still a strong force for the Norwegian industry, and most enterprises were still very committed to working for continuous improvement and reduced risk levels, with quite extensive funding being made available for risk reduction.

Climate issues were no apparent threat to the oil industry, and the prospects were to continue exploration and production for an unlimited period in the future.

New economic incentives (tax reductions) were introduced in Norway in 2005 in order to attract smaller enterprises and increase the number of enterprises operating in Norway. Several dozen enterprises were attracted, most of which focused on exploration, and if successful, often wanted to sell their prospects for a good price. About 20 years later, most of these enterprises have left the Norwegian industry or have been taken over by larger enterprises.

While the small enterprises without unlimited funding have mostly left the NCS, they started a shift in the industry that still has a strong influence on all enterprises, whereby the focus on profit has increased significantly, and the funding for improvement of health, environment and safety is significantly more restricted.

Equinor has taken over the petroleum activities of Norsk Hydro, and is in a monopoly situation in Norway, with 70–80% of the activities. There are two other medium-sized enterprises, AkerBP (owned by BP and Aker) and Vår Energy (mainly owned by Italian ENI). There are a few smaller European energy enterprises also with limited activities, one of them being Wintershall DEA.

In fact, Equinor, which is majority-owned by the Norwegian State, is the most aggressive operator and the least willing to provide funding for continuous HES improvement and risk reduction in its operations. A recent study (Bye et al. 2023) has documented this position in some detail, highlighting the negative effects for the working environment and safety in drilling operations, maintenance and modification, as well as in ISO (Isolation, scaffolding and painting) of production installations.

The decision-making approach described in the preceding sections requires an advanced level of trust between the parties, the petroleum enterprises, the employees and the regulatory authorities in order to function well. The prevalent framework conditions are now such that this trust is no longer there to the required extent.

This implies that several experts have started to argue that the approach to risk-informed decision-making ought to place more emphasis on the use of RTC, and that the regulatory authorities will need to formulate requirements that define what RTC are justifiable.

9.5 Questions

Q1: What are the respective advantages of a decision-making approach based on the evaluation of alternatives versus an approach based on RTC?

Q2: Why is trust important for the decision-making approach based on the evaluation of alternatives?

Q3: Please find all the current Havtil requirements for RTC and list all the required RTC that Norwegian operators have to concur with.

Q4: What is the relationship between quantitative and qualitative analysis when demonstrating that the ALARP has been reached?

Q5: Figure 9.2 shows that the uncertainty decreases during the execution of a project. It is still very advisable to focus on risk reduction in the early phases of a project. What are the underlying reasons for this?

References

Aven T, Vinnem JE (2005) On the use of risk acceptance criteria in the offshore petroleum industry. Reliab Eng Syst Saf 90:15–24

Aven T, Vinnem JE, Wiencke HS (2007) A decision framework for risk management, with application to the offshore oil and gas industry. Reliab Eng Syst Saf 92:433–448

Commission for Energy Regulation (2013) ALARP guidance-part of the petroleum safety framework. Version 2.0, CER/13/282

HSE (2001) Reducing risks, protecting people: HSE's decision-making process. UK Health & Safety Executive

ISO (2016) Offshore production installations—major accident hazard management during the design of new installations. ISO 17776. Geneva: International Standards Organization

Oil and Gas UK (2014) Guidance on risk related decision making, vol 2. The UK Oil and Gas Industry Association Limited

Standards Norway (2023) Risk and emergency preparedness assessment. NORSOK Z-013. Rev.4

Vinnem JE, Aven T (2006) Case illustration of a decision framework for health, environment, and safety management. J Risk Reliab 220:115–121

Tenth Lesson: Ethics in Risk Assessment

10

Ethics is knowing the difference between what you have a right to do and what is right to do.

—Potter Stuart

Abstract

Engineering ethics is an important but sensitive subject across all industries, ensuring the technical analyses align with societal values, justice, and long-term safety. This lesson concentrates on ethics in risk assessment, starting by elucidating its necessity. Then, the lesson discusses the ethical implications of risk analysis and risk tolerance criteria (RTC) in the planning processes of hazardous facilities, based on a Norwegian case study involving a liquefied natural gas (LNG) plant in populated areas. The Norwegian approach to risk-informed planning of hazardous facilities is outlined. Following this, the Norwegian case study and the main characteristics of the risk-informed decision-making in this case are presented. On this basis, the Norwegian approach to applying risk analysis and RTC is discussed, along with ethical challenges for enterprises, consultants, and authorities. Lastly, the potential misuse of risk analysis is discussed and highlighted. Among the ten lessons presented in this book, this final lesson holds the utmost importance, as it seeks to cultivate a heightened awareness among readers regarding their own ethical responsibilities and obligations.

X. Zhen and J.-E. Vinnem, *Ten Lessons on Risks*,
https://doi.org/10.1007/978-3-031-99064-9_10

10.1 Why Ethics in Risk Assessment?

In recent years, engineering ethics has become a widely discussed topic, the importance of which has received growing concern from the public, but not specifically for risk assessment. Actually, risk assessment is frequently perceived as being impartial. Some people even consider the risk assessment to be "objective".

The classical approach to uncertainty and probability is rooted in the premise that an "objective" or "true" probability exists. And that value can be found once unlimited time and resources are given to "repeat the tests". However, such ideal conditions are hard to find in practice.

In contrast, the Bayesian approach acknowledges the inherent uncertainty regarding future events and utilizes subjective probabilities to express the uncertainty with respect to what will occur. In the Bayesian approach, a "true" value does not exist, and the analysts' subjective evaluation of the uncertainty determines the results of the analysis. The authors concur with the views of many other scholars that risk assessment is inherently subjective, and the concept of an objective risk assessment does not exist.

In theory, the risk assessment could nevertheless be impartial in the sense that the analysts are neutral and do not favor any of the stakeholders. It would represent the analysts' subjective interpretation of the data, information, premises and assumptions, but without any bias towards either the enterprise or activity responsible, or the parties involved. This is an extremely demanding situation. For an individual to be completely neutral, the balance of the information needs to be maintained throughout the entire process. In comparison to the public in cases where risk exposure of the public from a hazardous facility is being debated, it is easy for the enterprise will gain more influence. The public involved will usually be a group of residents with their spokespersons, who may not be familiar with risk assessment and may be highly demanding in relation to the need for information, which may be experienced as extremely challenging by the analysts. The enterprise may be experienced as professional, factual oriented and easy to cope with in contrast to the representatives of the residents. Likes and dislikes are easily formed, implying that the completely unbiased analyst is a difficult role to maintain.

While completely unbiased risk assessments may exist in theory, it is unrealistic for most risk assessments to be impartial. In practice, completely unbiased risk assessments should perhaps be forgotten. As a matter of fact, the authors find that there is an increasing degree of ethical lapses of practitioners conducting risk assessment studies. It appears that risk analysts, to some degree, do not perceive themselves as impartial or unbiased at all, but more as consultants who can be "bought" to adopt the viewpoints of the clients. In this regard, the risk analysts could be comparable to some other management consultants, which are capable of adopting the viewpoints of their clients.

In addition, there are some new changes, which call for a greater emphasis on the ethical considerations in risk assessment. It is found by the authors that some enterprises have little or no experience in hazardous facilities management and the use of risk assessment. This implies that they will either perform simple risk assessment studies or commission such studies to consultants without relevant background knowledge. Further, the corresponding authorities do not have an extensive background in the use of risk assessment either.

In the face of these new changes, if all enterprises and consultants can behave conscientiously, there will not be a problem. If the responsible authorities can have an approach that is parallel with the approach taken by Havtil, there will be no significant problems either. When neither of these presumptions is fulfilled, there is a possibility of severe problems with societal safety. This is an emerging situation in recent years.

This lesson will discuss the important challenges of ethics in risk assessment based on a real case study of planning of hazardous facilities on the west coast of Norway (Vinnem 2010a, b). The importance of the real cases with respect to the ethics of risk assessment should be highlighted as they surely exist in practice, but few of these have been revealed to allow the industry and society to learn the lessons.

10.2 Norwegian Approach to Risk-Informed Planning of Hazardous Facilities

The Norwegian approach to risk-informed decision-making in the planning of hazardous facilities encompasses four primary components:

(a) Internal control principle,
(b) Use of RTC and ALARP,
(c) Use of risk analysis studies,
(d) Authority feedback and decision-making.

The major implications associated with each component are discussed in the following.

10.2.1 Internal Control Principle

The Health, Environment and Safety (HES) regulations in Norway rest mainly on the "internal control" principle (Hopkins and Hale 2002; Karlsen and Lindøe 2006; Lindøe and Braut 2009). Internal control means that enterprises are required to have a management system designed to ensure that all relevant requirements from regulatory authorities and other parties are adhered to and that this adherence is documented. Enterprises are thus responsible for adherence to laws and regulations and accordingly, it is not the duty of the regulatory authorities to accept or reject solutions adopted by the enterprises. However, regulatory authorities retain the right to conduct spot checks of adherence to regulations.

The majority of HES regulations are functional and risk-based. These regulations require that the risk level be tolerable, the requirements be functional, and they stipulate the functions to be complied with.

Within the Norwegian offshore petroleum sector, there is an extensive body of industry standards (the NORSOK standards) that have been developed through cooperation between the industry and the regulatory authorities. For more relevant information, readers could refer to Sect. 1.3 of the First Lesson in this book. These standards denote what the regulatory authorities consider to be accepted practice in fulfilling the functional regulations. For the onshore hazardous industry, there are no corresponding standards that define or formalize accepted industry practice.

10.2.2 Use of RTC and ALARP

The Norwegian approach to risk-informed decision-making in the planning of hazardous facilities is that the enterprises are responsible for the formulation of RTC, under the general obligation to choose solutions that ensure the risk level is ALARP, according to Section 20 of the "Act relating to the Prevention of Fire, Explosion and Accidents involving Hazardous Substances and the Fire Services" (hereinafter referred to as the "Fire and Explosion Prevention Act") (Norway Legislation 2002).

There are no guidelines or recommendations given by the regulatory authorities to assist enterprises in deciding which RTC to choose or at what level such criteria should be chosen. As a result, industry practice in this regard varies significantly.

It is not normal practice for Norwegian regulatory authorities to accept or reject the proposed RTC. However, there are clear differences with respect to how Norwegian authorities may give consent to solutions adopted on the basis of risk evaluations.

The lack of guidance is the same for the offshore petroleum industry as for onshore hazardous industries in Norway. For the offshore petroleum industry, most of the enterprises operating in the Norwegian sector have been active for 20–30 years and have a close relationship with the regulatory authorities, with the result that they are aware of what the regulatory authorities expect in most circumstances. Havtil, serving as the Norwegian regulatory authority for petroleum safety, also takes an active role in relation to the enterprises. It conducts studies and issues guidance with respect to the interpretation of regulations.

The Directorate for Civil Protection and Emergency Planning (abbreviated as DSB in Norwegian), which is the responsible authority for onshore hazardous facilities, does not have a correspondingly active dialogue with the enterprises. With a few exceptions, enterprises usually have no prior relationship with this authority.

10.2.3 Use of Risk Analysis Studies

The use of risk analysis studies in the planning and design of hazardous facilities is governed by the internal control principle, meaning that it is the responsibility of the enterprise to perform risk analysis studies and risk evaluation in a prudent manner and in accordance with applicable laws and regulations.

There is no approval or acceptance by regulatory authorities although they may ask to have a copy of the study for informational purposes. Regulatory authorities may also perform audits of the enterprise's management system.

In the Norwegian offshore industry, there is an extensive NORSOK Z-013 standard (Standards Norway 2023), containing requirements for conducting risk analysis studies. This standard is functional in the sense that the actual detailed scenarios to be analyzed and the data to be used are not specified. The NORSOK Z-013 standard is applicable to both offshore and onshore petroleum facilities that fall under petroleum law. It does not apply more generally to hazardous facilities onshore. It is of particular note that Havtil has commented on risk analysis in several cases where, in their view, the risk analysis was insufficient or unrealistic. Therefore, the petroleum industry is generally aware of Havtil's requirements in relation to the use of risk analysis.

For onshore hazardous industries, there is a general risk analysis standard, NS 5814 (Standards Norway 2008), which is applicable to all industries and societal activities and therefore has functional requirements on a very high level. However, there is no guidance available on which models, scenarios or data are to be used for hazardous facilities.

The regulatory authorities have sometimes subjected risk analysis studies to third-party verification, whereby a review is performed by independent experts.

10.2.4 Authority Feedback and Decision-Making

According to the Norwegian Fire and Explosion Prevention Act (Norway Legislation 2002), two clauses are of particular relevance to the planning of hazardous facilities:

§20. Level of safety in enterprises that handle hazardous substances, etc.

Enterprises are under an obligation to ensure that all handling of hazardous substances and dangerous goods is undertaken in such a manner that human life, the environment and surroundings are satisfactorily protected. Risk shall be reduced to a level which may be reasonably achieved.

§24. Gathering opinions from the public in the vicinity of enterprises with major accidents

During the planning of enterprises … …, such enterprises shall gather statements of opinion from the local public and take into consideration their comments regarding the siting and planned preparedness, response and safety measures … …

In the case study discussed in the next section, these requirements have been generally satisfied in a purely formalistic sense, meaning that the documents with the correct titles and headings have been prepared and submitted. However, a scrutiny of the contents of the documents and activities shows that this apparent compliance is sometimes significantly substandard or, to some extent, misleading. The response to §24, as quoted above, is typical in that reports were submitted and presentations made on the risk aspects of the LNG plant (in the case study), but only the risk to employees was considered. Given this limitation, the results must be seen as completely irrelevant in terms of risk exposure to the public at large.

There is a difference in practice between Havtil and DSB in respect to the feedback given to enterprises. Havtil has given such feedback regularly if the authority does not accept a particular risk study or its use for decision-making. In contrast, DSB does not appear to give similar feedback. Consent appears to be based mainly on satisfaction of formal requirements rather than on the applicability and comprehensiveness of the studies or the solutions adopted through the risk assessment process.

10.3 Case Study: Planning of an LNG Plant in an Urban Area

This case study is related to the development of an LNG plant in Risavika harbour, situated near the city of Stavanger along Norway's western coastline, as illustrated in Fig. 10.1.

The LNG plant receives gas from a gas processing plant at Kårstø, approximately 50 km away, via a subsea pipeline. The annual LNG production is expected

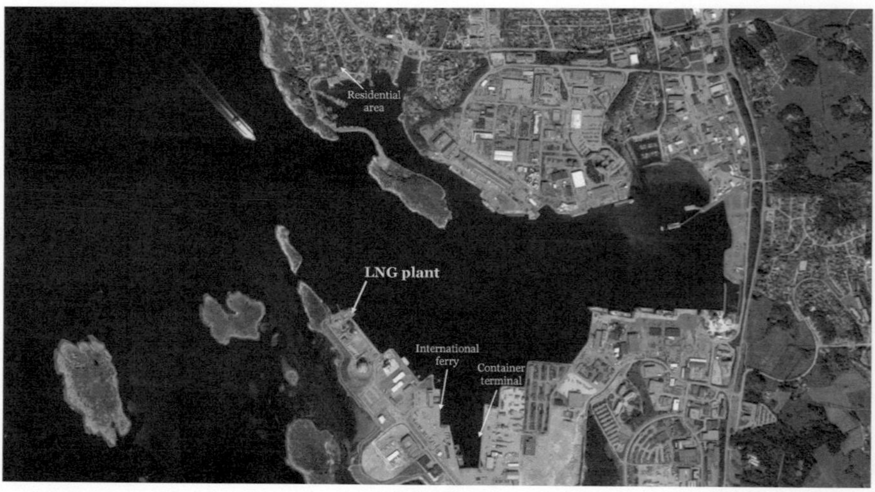

Fig. 10.1 Location of the LNG plant and its surroundings

to be 300,000 tons and can be increased to 600,000 tons should there be market demand. The LNG is exported to consumers via LNG tankers by sea or LNG lorries by road. The LNG plant mainly consists of the following elements:

- Pipeline landfall,
- Gas reception system,
- Gas pre-treatment system,
- LNG production system,
- LNG tank,
- Export system.

There are several other facilities around the Risavika Bay area. The following neighboring facilities are particularly important in that they generate substantial ship traffic, all of which has to pass by the quay of the LNG plant:

- Offshore supply base with extensive shipping traffic,
- Container terminal,
- Passenger cruise/ferry terminal.

The LNG plant was built and operated by an enterprise, which consists of 50% private ownership and 50% ownership by a group of local communities in the area around the plant. Nonetheless, the private ownership was terminated afterwards due to economic reasons of the project. The site is owned by an enterprise where local communities (some identical with the plant owners) have approximately 49% ownership. Particularly, local politicians are heavily involved as non-executive directors of all the enterprises involved. What is noteworthy is that all these enterprises act as if they were completely private enterprises without any restriction or obligations to the public, even when the local authority ownership is 100%. These issues must be seen as a severe deficiency in the applicable regulations. As a result, they violate the "open government" policies in Norway. For national and local administrations, there are significant requirements with respect to openness, in order to give the public the possibility to review the background to decisions made by these authorities.

The most significant issues related to the approval process of the LNG plant are summarized in the following (Vatn et al. 2008):

- In June 2006, the planning of the LNG plant was approved by the local authorities. A series of QRAs was carried out, but the risk level to third parties (societal risk) was not addressed.
- In December 2007, the authorization to store and treat inflammable goods was given by DSB on the basis of a preliminary risk analysis (PRA). The PRA was conducted based on average values in relation to both reliability data and layout of the plant, since the detailed engineering had not yet been completed. The analysis did not address risk exposure to neighbours of the LNG plant.

- Updated risk analyses, which need to show that the risk level of the LNG plant as built meets the acceptance criteria, have to be provided so as to acquire the final authorization from DSB. Test production was expected to start in 2010.

It can be seen that the detailed engineering of the plant ran in parallel with the safety approval process, which indicates that the initial safety approval needs to be issued on the basis of the PRA.

The great concern of this case study is drawn on the early planning process, which is used to select the location of the LNG plant as well as consider its risk exposure to local communities. From a societal safety point of view, the decision on the selection of such a location does not appear to be logical. Though there had been significant resistance against the planning of the LNG plant among its neighbors, the construction project was approved by the local authority and DSB. In 2008, this conflict evolved into a big public issue in the media. The initial conflict focused on the risk communication, which was not properly conducted, while the substantial criticism was raised to the quality of the risk analyses as well as the decision-making process related to societal safety (Vatn et al. 2008).

In this case, the use of risk analysis studies in the risk-informed decision-making for the LNG plant in the early project phases should have focused on the following aspects, which should have been investigated and discussed.

- Choice between alternative locations,
- Risk analysis in tolerability evaluations,
- Risk communication with the public,
- Use of risk analysis in the design process.

10.3.1 Choice Between Alternative Locations

Initially, two potential sites were considered for the LNG plant: Risavika and Mekjarvik. These two locations are approximately 10 km apart. Nonetheless, this decision-making process was carried out without any public input, in spite of the extensive involvement of local councils, who are the owners of the enterprise, and despite explicit requirements that such information should be communicated in the public domain. The main characteristics of these two locations are presented and compared in the following:

Location 1: Risavika (chosen site)

- Close to urban area.
- Close to several other public and industrial facilities, including an international ferry terminal.
- Jetty located in the mouth of a long bay with extensive passing traffic of large ships.

- Unfavorable direction of prevailing wind with respect to densely populated areas, such as the international ferry terminal.

Location 2: Mekjarvik (alternative site)

- Not an urban area.
- Easy to separate marine traffic from jetty location.
- Favorable topography that restricts the spread of a gas cloud towards residential areas, directing it instead over the open sea.
- Favorable direction of prevailing wind with respect to densely populated areas.

It can be seen that these two locations are quite different, especially from a safety point of view. It was found that in the process of choosing the location of the LNG plant, no risk analysis was performed at that stage and no risk evaluation was made. The decision appears to have been made merely in consideration of project economy and market conditions, as risk evaluations would probably have shown that Location 2 was significantly better from the point of view of major accident risk.

In comparison to the location of petroleum facilities, which fall under Norwegian petroleum law, risk evaluations are compulsory activities in all stages of planning in accordance with the Management Regulations (Havtil 2020). There are rigid requirements that ensure public knowledge and involvement in the evaluation of premises and decision-making at certain stages. In this sense, the administration of the Norwegian Fire and Explosion Prevention Act has severe deficiencies with respect to ensuring that safety is integrated into early planning and screening processes.

10.3.2 Risk Analysis in Tolerability Evaluations

The risk analysis results for the LNG plant from both the initial QRA study (Capra et al. 2007) and the QRA study during the detailed engineering phase (Linde Engineering 2008) are shown in Fig. 10.2. The upper and lower limits of risk tolerability are also presented.

In comparison to the risk analysis results in the initial QRA study, it can be seen that the expected risk level in the updated QRA study has increased by a significant margin. For instance, the updated frequency per year of accidents with at least 100 fatalities was increased by a factor of 56. This further indicates one of the main problems in the present case, i.e., lack of robustness in the initial QRA study. What calls for special attention is that the initial QRA study had been conducted before the detailed engineering studies started, while the updated QRA study was performed by the engineering contractor, reflecting all the engineering details. The difference between the risk analysis results of the initial QRA study and those of the detailed engineering phase could be expected if the initial QRA

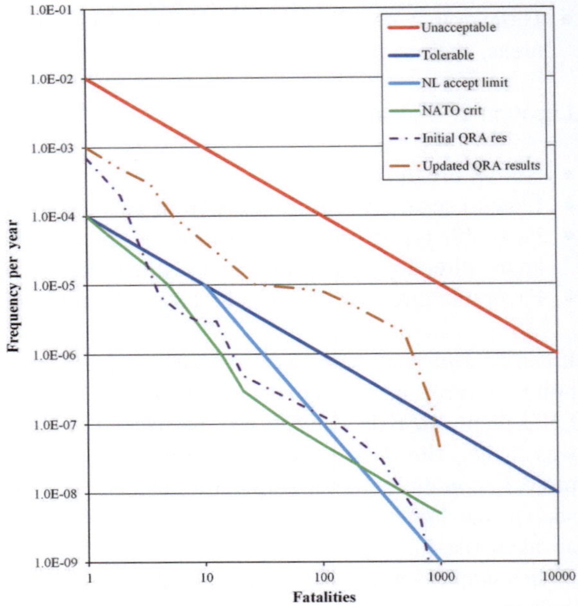

Fig. 10.2 Updated risk analysis results for the LNG plant and alternative risk tolerance criteria. Adapted from Vinnem (2010a) with Elsevier permission

study was coarse. Nonetheless, this difference should have been recognized and due allowance should be made for uncertainties.

The risk tolerance limits, which were proposed by HSE, were adopted by the operator. Thus, the initial QRA results were interpreted as "negligible". Notably, these risk tolerance limits were intended for use with respect to existing facilities. More strict limits, where risk aversion has been incorporated, have been stated for new facilities by HSE (2008). In addition, Fig. 10.2 presents other risk tolerance limits from the Netherlands (NL) and North Atlantic Treaty Organization (NATO), which may also be used for a broader illustration of the risk tolerability.

10.3.3 Risk Communication with the Public

In the present case, one major problem is the communication of risk analysis results to the public. The initial QRA results were not available until one year after the public hearing of the planning proposal. For the input to the public hearing, the plant developer carried out the coarse PRA-type studies, where risk analysis results were presented in a risk matrix, as shown in Fig. 10.3.

In this risk matrix, the most severe consequence category was "critical", which was indicated as "more than one fatality". Nonetheless, this type of risk presentation is highly unsuitable for the characterization of the risk exposure posed by major hazards to third parties, as it only provides a fragmented view of each hazard without offering a comprehensive assessment of total risk. Further, although the risk analysis was limited to risk to employees on the LNG plant, these risk analysis

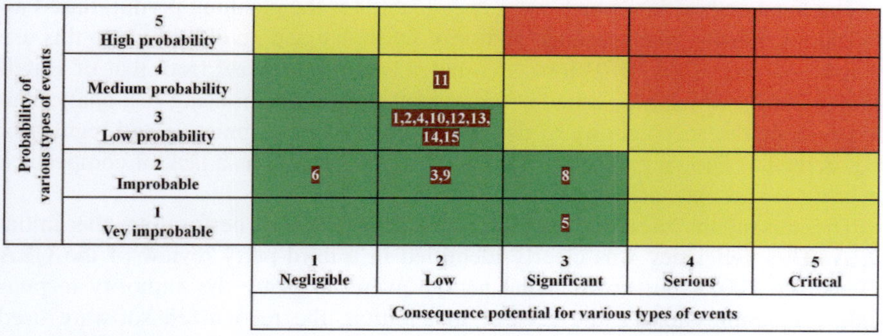

Fig. 10.3 Risk picture for the LNG plant and export facilities. Adapted from Vinnem (2012) with Elsevier permission

results were used by the enterprise in the compulsory communication with neighbors to present their potential risk exposure. Thus, a couple of complaints were made to the enterprise and all involved authorities for the effect that employee risk was completely unrepresentative of the risk to the public. However, none of the parties acknowledged these concerns. The reason is hard to comprehend, but the most obvious interpretation is a severe lack of competence.

10.3.4 Use of Risk Analysis in the Design Process

In several risk analysis studies of the LNG plant, one key assumption was made that the gas release would be ignited immediately in the event of impact of a passing vessel on an LNG tanker loading at the quay, presumably by sparks generated by the collision itself. Nonetheless, there was no explanation of how such ignition of a very heavy and cold gas could occur physically. Actually, it is hard to foresee how ignition could occur in this way. However, the implication and purpose of this assumption in the studies were that it was unnecessary to consider any spread of the gas cloud due to wind and heating of the liquefied gas, with obvious consequences for the scenarios that the public might be exposed to. This is a very critical assumption, which should have been subjected to a sensitivity study in order to illustrate how changes in the assumption would affect the results and the robustness of the assumption.

In this case, the risk analysis was used as a tool to "prove" that the LNG plant was safe enough not to follow the US practice for safety zones. The analysis rested on arbitrary assumptions about leak rates, leak durations and an optimistic assumption that all safety systems would always function as intended. The result of this approach was that a safety zone, which was regarded as dramatically insufficient by many professionals, was adopted. Confirmation of its insufficiency came from the updated QRA study, which showed that the risk level for the public had increased considerably, as shown in Fig. 10.2.

The local authorities, which were responsible for the planning permit, raised no objections. Further, the specialist authority DSB also had no objections to this use of risk analysis. Thus, DSB took a completely different view from that of Havtil (Havtil 2007), which had issued a warning of malpractice in using risk analysis to demonstrate the acceptability of deviations from accepted practice and regulatory requirements. The view taken by DSB may be interpreted as a lack of competence or professional maturity.

The consultant used the unrecognized software for performing the initial QRA. This deficiency was clearly identified in a third-party review of the QRA (Danielsen 2007), but no remedial actions were taken by the authority responsible for commissioning the review. In addition, the recognized software used in the updated QRA was not suitable for the detailed engineering stage. At this stage, more detailed modelling based on computational fluid dynamics is always expected. Good practice would prefer to use several models in order to demonstrate the robustness of the QRA results.

The safety zone of the LNG plant was determined on the basis of a specific leak scenario with a short duration of merely 30 s. Such a short duration assumes that all relevant safety systems (barriers) would function exactly as intended with no delay at all. However, leak statistics from both onshore and offshore petroleum facilities have demonstrated clearly that gas leak cases rarely occur exactly as "planned". Delays in detection or failures of automatic isolation requiring manual isolation are more common than rare, frequently resulting in significantly longer leak durations.

The approach taken in this case may be characterized as non-conservative. No sensitivity studies were made, nor was the question of possible variations addressed. Therefore, the safety zone was considered an unrealistic minimum case with a radius of 125 m. It is worth noting that the US authorities do not accept the use of risk analysis in the definition of safety zones and instead stipulate a safety zone radius of up to about 1600 m.

A robust definition of safety zones should probably be based on a leak scenario with a duration of several minutes, which should reflect situations where one or a few barriers fail or where a delayed response occurs. Of similar importance would be the performance of sensitivity studies to assess the effect of changes in assumptions and premises in relation to the dimensions of the safety zones.

In the case of the LNG plant, the initial QRA study had made assumptions about the explosion strength of the closest buildings outside the safety zone, but the local authorities had not ensured that this had been followed up on in the construction of the building in question. A contributory factor was the poor documentation of the initial QRA study, making identification of what should apply far from obvious.

10.3.5 Cause of Serious Deficiencies

This actual case study has shown several serious failures of competence on the part of the enterprise as well as the local and national authorities. Some would argue that the failures are so conspicuous that the parties involved appear not to have absorbed any of the relevant professional experience.

The current Norwegian legislation is aimed at fulfilling the EU Seveso II directive in a prudent manner and ensuring that the public is not compelled to accept an unreasonable level of risk exposure. In the present case, the public was obliged to accept this unreasonable risk exposure.

There are several societal control systems, such as consultation with the public, politicians' consultation with the enterprise during the application process for building permit, etc., which have been designed to prevent third parties from being subjected to significant risk levels without being consulted and to ensure that their views are taken seriously. All these control systems failed in the present case, with the result that the public was obliged to accept the unreasonable level of risk exposure. There was one complicating factor, which was that the highest risk exposure would apply to passengers in an international ferry terminal. Obviously, these passengers do not have any spokespersons and their voices were not heard in the process. The owner of the land on which the ferry terminal has been built is also one of the owners of the LNG plant. Thus, the owner of the land is not an appropriate person to represent the passengers in the terminal in an objective way. In theory, the ferry company should be the spokespersons of these passengers, but in practice, the ferry company did not do this.

The most surprising and worrying aspect of this case is that both local and national authorities failed to realize that the obligations under current laws and regulations to allow the neighbors a substantial say in the issues had not been observed. One of the arguments in the present case focused on the assumption by the authorities that they could not all be wrong and thus who objected (including the author) had to be wrong, apparently because the majority opinion reflected the opposite point of view.

Nonetheless, it is certain about the fact that qualitative analysis for the evaluation of the risk exposure of the third party as well as the determination of safety zones, is highly insufficient. The practice in the Netherlands (Uijt de Haag, et al. 2008) also confirms this deficiency. It is also certain that the employee risk is not representative for being used to inform the public with respect to their risk exposure. The lack of satisfactory involvement of the public becomes a serious failure of societal control systems.

In an industry, it is essential that the enterprises and regulatory authorities take the obligation seriously to protect the public as a third party, whereas it is a pity that both the enterprise and the regulatory authorities in this case failed to assume this responsibility.

10.4 Norwegian Approach to Applying Risk Analysis and Risk Tolerance Criteria

On the basis of the preceding case study, this section further discusses the Norwegian approach regarding risk analysis and RTC, encompassing execution of risk analysis studies, the use of RTC, Havtil considering changes to RTC, a comparative analysis of Norwegian and international perspectives, and risk communication to the public.

10.4.1 Execution of Risk Analysis Studies

The Norwegian practice is that the enterprises are responsible to the task of conducting risk analysis studies under the internal control principle. Havtil has made extensive reference to the NORSOK Z-013 standard in its guidelines to the regulations and has issued requirements implicitly for the execution of risk analysis studies.

For the land-based hazardous facilities under the Fire and Explosion Prevention Act, there are no specific requirements at all, merely the obligation to carry out risk analysis. The coarse PHA (Preliminary Hazard Analysis)-type risk analysis was even accepted by DSB as suitable. DSB had an independent verification made for the initial QRA study, the significant weaknesses of which had been found, but DSB failed to ensure that these weaknesses were rectified.

The Dutch authority has provided a solution in the following manner (Uijt de Haag et al. 2008):

> Risk analysis studies for safety zone definitions must be based on the use of predefined, recognized software with predefined scenarios and assumptions.

For a given scenario, this solution will reproduce identical results each time, irrespective of which consultant performs the risk analysis studies. This can be achieved if the scenario is well-defined and repeated frequently. If the scenario is variable, so that detailed, predefined scenarios and assumptions cannot be made, the alternative in defining the scenarios would be to state principles to be applied. For instance, these principles may be related to:

- Leak scenario: full flow in main piping for tanker loading;
- Leak duration: specified time until isolation is completed;
- Specified frequency for defined leak scenarios;
- Defined environmental conditions;
- Definition of scenario with respect to affected individuals.

Based on the experience from the present case study, it is obvious that the Dutch approach is a distinctly more robust and defensive way of addressing risk-informed decision-making and planning of hazardous facilities, in comparison to the Norwegian practice. The Norwegian approach may be acceptable when the regulatory authority takes an active role in reviewing the QRA studies to ensure that the use of risk analysis studies is defensible and justifiable.

It may be questionable in the present case whether the severe faults in the risk analysis studies could be attributed to a lack of knowledge within the companies that performed the risk analysis studies. In part, it may be so due to their limited experience with such studies. Nonetheless, an international consultant was contracted for the initial QRA study, apparently due to its extensive experience. This implies that one may question whether lack of experience was the only factor, while lack of ethical standards may also have contributed to the faults. When very critical assumptions are virtually undisclosed, this is most likely not a question of the lack of knowledge, but a lack of openness. If this cannot be achieved, the Dutch approach should be adopted. Principles for how to define scenarios and premises should be included in the Norwegian approach.

If the analysis of a facility is built on assumptions and premises regarding facilities outside the safety zone, it will obviously be of vital importance that this be communicated to the affected facilities. Assumptions and premises may imply restrictions on activities and may also have implications for the dimensioning of structures and equipment in the neighborhood.

Such a challenge quite often arises in the case of production facilities in the Norwegian offshore petroleum industry. In several accident and incident investigations (Schiefloe 2005; Jacobsen et al. 2008), Havtil has found this to be one of the contributing causes. Follow-up of assumptions and premises is actually an element given considerable weight in the petroleum industry. It has been found in several cases that when modifications or operations are planned for a facility in the operational phase, the responsible persons often forget to check if premises and assumptions in the risk analysis have implications for the work in question. This may affect the actual execution of the maintenance work but may also apply to the inhibition of safety barriers, and the extent of simultaneous activities.

10.4.2 Use of Risk Tolerance Criteria

In the present case study, the RTC had been defined by the enterprise, under the internal control obligation according to Norwegian regulations. The national authorities do not publish any guidelines with respect to the RTC except for the obligation in Section 20 of the Fire and Explosion Prevention Act to reduce risk as far as reasonably practicable.

Hence, in addition to a narrowly focused comparison between results and criteria, it would be expected that broad evaluations should be performed in order to elucidate the implications of the evaluations, the robustness and other implications, not least for emergency preparedness. The enterprise should conduct this work and supply the relevant information. If not, it should at least be the duty of the national authority to supply such information, especially when the neighbors address formal complaints to the Ministry. None of this occurred in the present case. The Ministry of Justice and Public Security and DSB referred to the evaluation made by the enterprise, but even the implications of the weakness found by the independent verification were not considered in relation to the formal complaint.

Additional evaluations and clarifications are particularly to be expected before those responsible determine that the risk level is negligible. In the present case, all involved parties arrived at a conclusion of negligible risk without any additional evaluations. Even the independent research team, which was hired by the enterprise to improve risk communication, stated that the risk level was considered to be negligible (Vatn et al. 2008), without consideration of robustness and validity. The failure/implication in this attitude was particularly emphasized by the enterprise's own updated QRA study, which showed that the frequency of accidents with at least 100 fatalities increased by a factor of 56 in comparison to the results from the initial QRA study.

In addition, the basis for the evaluation of risk is also a sensitive issue. The evaluation in this lesson has referred to the RTC used by Dutch authorities for similar petroleum-based facilities (Uijt de Haag et al. 2008), as well as criteria adopted by NATO for the location of ammunition depots in populated areas (NATO/PFP UNCLASSIFIED-III-1-1 2003), as shown in Fig. 10.2. Both these sets of RTC are stricter than those used for the present case study, which are the same as those used by HSE in the UK. One of the main differences between the adopted risk tolerance limits of enterprises and these alternatives is that the Dutch and NATO criteria have included risk aversion, whereas the criteria adopted by the enterprise do not.

In comparison to the alternative RTC, the initial risk analysis results of the present case are close to, sometimes slightly above, the risk tolerance limits. Increases due to changes in assumptions would give results significantly above the acceptance limits, as confirmed by the updated risk analysis results.

There is no tradition in Norway for the regulatory authorities to formulate an RTC for the industry to use. The enterprises themselves are required to develop their own RTC when applying for permits for new major hazard installations, such as offshore and onshore petroleum facilities, as well as relevant major hazard plants onshore. Though there is no tradition for the regulatory authorities to formulate any guidelines, Havtil has influenced petroleum enterprises operating on the NCS to the extent that the RTC used are fairly similar, if not identical. This alignment has been relatively easy to achieve, because there are approximately ten enterprises involved in these offshore and onshore operations under petroleum law and most of them have been operating in Norway for many years. This approach has proven to work reasonably well for operations under petroleum law.

The practice under the Fire and Explosion Prevention Act is quite different. Few of the enterprises will have any experience with major hazard management. There will normally be new enterprises for each new project and the national authorities will also be relatively inexperienced in these matters because several years normally elapse between each new project with major hazard potential.

A general principle for the use of risk analysis is that its scope should be to identify how risk may be reduced through sensible risk-reducing measures, rather than to demonstrate that deviations from accepted practice are defensible. When the offshore petroleum industry in Norway started to use risk analysis in the early 1980s, this was one of the most common misuses. This practice was curtailed after some time, but has recently become more of an issue owing to the advent of newcomers to the offshore industry.

Why is it not justifiable to use risk analysis to demonstrate that deviations from accepted practice are acceptable? This is related to several factors, one of which is the robustness of risk analysis. The use of risk analysis is often open to possible manipulation by the user, and even subtle changes in assumptions may have a dramatic influence on the results, e.g., for third parties. Another aspect is that risk analysis is often too insensitive to possible deviations. This reflects a level of coarseness in the models that could be avoided if adequately addressed.

Finally, there is a wider debate about RTC and ALARP. One of the principles (Aven and Kristensen 2005; Aven and Vinnem 2007) is that RTC cannot be used in isolation for the determination of the tolerability of risk. However, the use of RTC is quite widespread in the industry and any discussion of risk communication practices has to relate to this fact.

10.4.3 Havtil Considering Changes to Risk Tolerance Criteria

The use of RTC in Havtil has formally been part of legislation since 1991, after having been an informal requirement since 1981. It has thus had a high profile in Havtil's communication. It should be noted that there is also a requirement in the Norwegian petroleum legislation for continuous improvement for all HES aspects. But in the last 10–15 years the enterprises have kept their risk tolerance limits constant, in spite of nearly tenfold improvement of the fatality statistics in offshore operations. A FAR level of 10 which was an ambitious goal in the mid-1990s, is virtually without effect at all today.

As a result, Havtil has started recently to consider if risk tolerance limit requirements shall remain as part of legislation, after having been criticized for ignoring the fact that no improvement of risk tolerance limits occurred. It is important that Havtil recognizes that such decisions are extremely vital and they are prepared to not rush into decisions.

Four cases are considered as illustrated in Fig. 10.4, from removing all limits and just using ALARP, to the extension of the regulatory requirements. So far, there has been no decision made by Havtil, even though it has been considered for a few years.

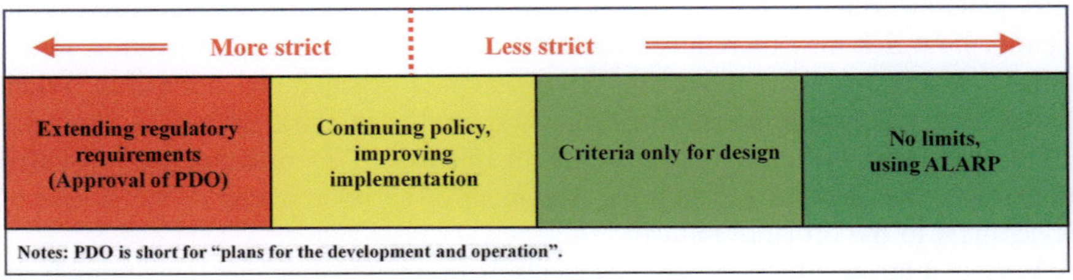

Fig. 10.4 Possible changes considered by Havtil for changes to RTC requirements

10.4.4 Norwegian versus International Perspectives on Risk Analysis

The Norwegian approach, as outlined in Sect. 10.2, has been based on good experience from the offshore sector, where risk analysis has been used well for over 40 years. In the initial period, the regulatory authorities suggested acceptance levels (in the form of maximum frequency of events not included in the design basis), but for almost 30 years, it has been left to the enterprises to define their own RTC. Up to now, this Norwegian approach to the use of risk analysis and RTC is considered to function well.

In the present case study, the use of risk analysis and RTC for land-based hazardous facilities under the "Seveso regime" has been demonstrated to fail severely. This is an experience widely at odds with that from the offshore petroleum sector. One of the main building blocks in the Norwegian approach is the internal control system, whereby enterprises are granted permission to choose their own RTC and to plan and carry out the risk analysis studies. The extent to which the regulatory authorities choose to involve themselves in the auditing of the enterprises' performance of the studies and evaluations under the internal control obligations is variable. Havtil has a practice of active involvement as well as in-depth audits and reviews. DSB apparently has a practice which is restricted to formalistic verification that the required documents have been prepared, with limited interest in the actual professional contents and the solutions adopted. This difference may well be the main determining factor with respect to the success of the Norwegian approach.

A different model can be found in some other countries and regions. For instance, the Netherlands has had risk tolerance criteria in place for more than 20 years, since the publication of the policy document "Premises for Risk Management" in 1988–1989. The Dutch approach has included risk aversion in the risk tolerance limits defined. Up to now, the UK, Switzerland, Canada and Australia have policies similar to those in the Netherlands. In addition, the Dutch authorities have found the need to modify the procedure used for planning purposes in order to make the model more rational and robust.

One aspect where risk-based decision-making differs between offshore petroleum installations and onshore hazardous facilities is that offshore petroleum installations do not expose any third party to risk by their nature. Risk exposure is limited to employees, entailing a significantly simpler decision-making process. For this reason, there is less focus on risk aversion.

In the light of positive experience from the Netherlands and other countries, the approach whereby the regulatory authorities stipulate acceptance limits appears to be more successful in comparison with the Norwegian approach based on practice from the offshore petroleum industry where the formulation of acceptance limits is left to the industry. The present case study certainly demonstrates that the Norwegian approach has failed in a number of ways in the actual circumstances.

In theory, the regulatory authorities should intervene if unreasonable acceptance limits are chosen by an enterprise. Practice demonstrates that intervention is unlikely, including in the case of offshore petroleum activities as Havtil also does not make any intervention. Given the lack of clout demonstrated by the national authorities in the present case study, it is quite unlikely that any intervention would have occurred, and it certainly did not.

It appears that no other countries allow the specification of safety zone dimensions purely based on risk analysis studies without restrictions or guidance on the degrees of freedom that the enterprise may take.

10.4.5 Risk Communication to the Public

In this case, the risk analysis results, which were presented to the public to illustrate their exposure, took the form of a risk matrix (refer to Fig. 10.3). The risk matrix is not able to differentiate between on-site (LNG plant) and off-site (the neighbors of the LNG plant) effects, and the analysis had merely considered on-site effects. Our recommendation is that a risk curve similar to that in Fig. 10.2 should replace the risk matrix. We may further ask, would a risk curve like Fig. 10.2, if presented at the time of the hearing, have been sufficient to provide suitable risk communication with the public? The reply to this is that a mathematical expression needs to be supplemented by an explanation of the practical implications of the results, discussions of uncertainties and sensitivities, illustration of alternative perspectives, etc.

Some may argue that it is not possible to carry out a quantitative study at an early planning stage because only limited technical information and drawings are available. But the authors do not agree with this view. The main volumes, capacities of main equipment units, as well as distances and directions, are all known. This means that it is feasible to carry out a quantitative study in relation to third party risk. The Dutch experience (Uijt de Haag et al. 2008) also indicates that quantitative analysis is used in the early planning stage. On the other hand, it may not be possible to analyze employee risk quantitatively, due to lack of layout drawings, and other necessary technical information.

Contributions to individual risk levels should be presented in order to reflect the relative impact of different plant sections and sources. It would also be valuable to present risk contributions from the different accident mechanisms, such as fire, explosion, and the timing of ignition.

Several types of sensitivity studies should be performed, one of which should be a study where parameters and assumptions in the risk model are varied to determine effects on the risk analysis results. This is essential in order to determine how robust the risk model is and how changes in premises and assumptions may influence results and conclusions. The most important type of sensitivity studies should focus on demonstrating the effects on risk results from potential risk-reducing actions. In the present case, no sensitivity studies were conducted in the initial QRA study but a few were performed in the updated QRA study, relating to a limited set of proposals, such as:

- Installing a rock wall around the LNG storage tank,
- Reducing the size of cooling medium tanks.

All the risk reduction proposals were discussed qualitatively. The rock wall had a significant quantitative effect and was proposed for implementation. The second proposal was argued not to have a positive effect, possibly the opposite, due to the increase in the number of potential leak sources.

What was lacking from this evaluation of risk-reducing measures was some risk reduction proposals which would potentially reduce third-party risk overall in a significant manner, but would have a clear negative effect on the operation of the LNG plant. Such proposals could include:

- No shipping traffic permitted during LNG loading operations;
- No LNG loading permitted while the passenger terminal is in use by ferry passengers.

Given the attitude demonstrated by the enterprise in this case, it is hardly surprising that restrictions of this kind have not been considered.

10.5 Ethical Challenges for Enterprises, Consultants, and Authorities

The planning case of the LNG plant in an urban area has shown the serious failures and deficiencies on the part of the enterprise, the consultant, and the regulatory authorities. Among these, the ethical challenges in risk assessments stem from the fact that decisions regarding risk tolerability and plant requirements are entirely based on risk assessments, which are planned, executed, documented and approved without any form of official evaluation, control or acceptance under the internal control principle. The choice of analytical approach as well as the timing of studies and evaluations also falls under the internal control principle.

The expectation of the performance of risk assessments is that they shall be completely neutral. However, the misconception that risk assessments could be "objective" may still be a common assumption favored by regulatory authorities. It is beyond doubt that the results of risk assessments may be manipulated severely through changes or omission of assumptions and premises. A thorough appraisal of assumptions and premises is virtually impossible to implement in practice without redoing the analysis. This problem could actually be improved somewhat if the documentation from risk analysis studies were more comprehensive than what is the norm. If the consultants were very careful to include a thorough discussion of all critical assumptions and premises, the review of assumptions and premises would be made much easier. Nonetheless, such documentation is rarely seen in the current risk assessment practice.

The public is usually much less experienced with risk assessments than other stakeholders. This implies that it will usually be much more demanding for the public to determine if the enterprise is fulfilling its obligations under the relevant laws and regulations. The public often finds it difficult to form an informed opinion due to factors, such as the choice of the analytical approach, the data used, and the timing of the analysis.

In this book, the focus of ethical challenges is on consultants and the enterprises (clients). These ethical challenges are discussed respectively in the following.

10.5.1 Ethical Challenges for Enterprises

When hazardous facilities shall be planned and designed, the obligation of the enterprises is to select solutions and locations that ensure a risk level as low as reasonably practicable. Yet it has been repeatedly seen over the years that enterprises make a choice between alternative locations for new facilities based on economic consideration merely, without any consideration of safety aspects. The underlying assumption is that safety is something that "can be fixed" afterwards.

If the plant to be built is a facility without possible accidental consequences for the public, such as a mechanical workshop or a dairy production unit, there may not be any problem with the "fix afterwards" approach. However, if it is an ammunition factory or a warehouse for flammable or explosive chemicals storage, it will obviously be very crucial that the location in relation to third parties is carefully considered.

Under the Norwegian regulations for implementation of the Seveso II EU Directive, the obligation is on the enterprise to conduct the public hearing with all stakeholders about the risk level and the emergency preparedness arrangements that are planned. This is a prerequisite for the enterprise in order to obtain planning permission. The information submitted for the public hearing is completely controlled by the enterprise. The enterprise bears the sole responsibility for presenting truthful and correct information. This reflects a basic aspect of the Norwegian societal system, which is largely trust-based.

The local authorities will be informed about the public hearing, but they cannot be expected to have expertise in risk assessment nor to be able to assess whether the information is adequate according to the requirements. When the enterprise claims that the information is adequate and that it has fulfilled all their obligations, no one is able to argue against that claim. Consequently, the planning permission will be given by the local authorities on false premises.

The risk picture is also presented by the enterprise and its consultants, without giving others the opportunity to raise objections on a professional basis. In the present case study, the information presented by the enterprise was that the plant would pose zero risk to the public, in spite of the fact that the initial QRA study documented scenarios with up to 1000 fatalities occurring with a significant probability, certainly not a negligible risk as the enterprise and its consultants attempted to portray.

The presentation of risk in the present case study appears to be like wearing "blinkers" and is very narrow-sighted in an "as-little-as-you-need-to-know" manner. The impression is that it was intended to present the risk level to the public like it was negligible without informing them of alternative conclusions that could be reached with altered premises. The information was only considered on the basis of the RTC chosen by the enterprise, which are actually less stringent criteria than those used by other enterprises. No information was provided on the stricter criteria applied in other countries, nor about other Norwegian enterprises using stricter criteria. The RTC were referenced to HSE guidelines in the UK, but the information from HSE stating that these criteria are not applicable to new facilities was not provided.

In a proactive assessment of the feasibility of the proposed new facilities, it is expected that broad evaluations are presented in order to illustrate the tolerability of risk in alternative ways.

It has been documented that alternative criteria would have stopped the project, especially after the revised risk picture was presented by the updated QRA.

A potentially serious unethical behavior of an enterprise is that it affects the choice of premises and assumptions used by the QRA studies in order to misrepresent the risk picture. In several cases, the suspicion was that this unethical behavior had occurred, but it is hard to prove.

10.5.2 Ethical Challenges for Consultants

The ethical demands on the risk analysis contractors are significant, and there are normally two types of unethical behavior:

(1) If a risk analysis contractor is prevented by the client from carrying out a risk assessment with a scope of work that complies with the applicable regulations and legislations, does the risk analysis contractor have a responsibility to rectify this situation or inform the regulatory authorities, or is it just fine to accept the client's instructions? This is an ethical challenge that the risk analysis contractor for the LNG plant failed to meet in a responsible manner. The result was an unethical risk analysis study, which the client used as if it complied with the relevant regulations and legislations.

(2) Is it ethical for a risk analysis contractor to please the client by hiding the crucial assumptions that may change the results dramatically in an appendix that the typical readers will never discover it? The right answer is obvious. However, this is precisely what has been done in the LNG case with the assumption that all leaks from the tankers would be ignited immediately. Not only is it a wrong assumption, but it was hidden and not open for discussion and evaluation.

The first unethical behavior may have been made unintentionally based on the client's insistence and the contractor's insufficient resistance to pressure. However, the second unethical behavior gives the impression that it has been done deliberately, which will be even more serious. The second unethical behavior is further discussed in detail in the following.

In the present days, it is relatively common in advanced QRA studies that significant computer modelling is an integrated part of work. This implies that all details will not be available through the reporting format, but will have to be found inside the computer models, which are usually not available for review. The effect of this is that a lot of information, such as some assumptions, premises and data, will not be documented. The risk analysts will have to be trusted to behave ethically and impartially. It is normal to assume that this is the case. Nonetheless, over the years, some cases can be seen from time to time, where unethical behavior of risk analysts is a concern, especially if the case has implications for the public and there is no rectification made by regulatory authorities or anybody else.

There is one relatively common fault, i.e., portraying the risk analysis results to be much more robust than they actually are. This will be referred to as unethical only if it is a serious deviation between what is done and what is presented.

The severe fault is to adopt unrepresentative assumptions, which exclude critical scenarios. This may be a simplification, which could be defensible or not, depending on the circumstances. This fault becomes extremely serious if the assumptions are not addressed when presenting the results. It is expected that the implications of such assumptions on the results are discussed, together with a discussion of possible changes to the results if alternative assumptions were adopted. If all of these elements are lacking, unethical behavior may arise, characterized by attempts to conceal what has been done.

In the planning case of the LNG plant in an urban area, one of the early PHA studies adopted the assumption that all leaks from the tanker would be ignited immediately, thus eliminating the possibility of drifting cold LNG clouds over water, without stating in a visible manner that this assumption was made, nor illustrating the effects on the results of alternative assumptions. It is difficult to imagine how this assumption could be adopted in the first place, or how one could assume that a cold, dense cloud could be ignited immediately, given that it would presumably be in the liquid state or at 100% gas concentration. As such, this assumption appears wrong, but this is beyond the point. The issue is that the the studies created the impression of attempting to conceal what was assumed, and certainly failed to illustrate the effects of alternatives.

Another form of misrepresentation is to draw conclusions about aspects that were not included in the modelling. In the planning case of the LNG plant in an urban area, one of the later QRA studies made the conclusion that no scenario had been identified in which gas clouds could travel over water and reach domestic areas. Nonetheless, this analysis had adopted the results of the PHA study, which had ruled out drifting gas clouds due to its assumption with respect to immediate ignition. Essentially, the QRA study had adopted the same assumption as the PHA study. Therefore, it becomes a severe case of misrepresentation to draw conclusions about the absence of gas clouds without documenting this crucial and probably erroneous assumption with respect to immediate ignition.

Another case of misrepresentation is to exclude significant parts of the experience data from the analysis on the grounds of not being applicable. This may be justifiable and actually good practice if there are valid reasons for excluding the events in question. Similarly, failing to exclude data that should prudently be excluded is also a failure. Both these two actions may be good risk assessment practice, as long as it is done openly and supplemented by a discussion of the effects of the assumptions that have been adopted and the consequences of making alternative assumptions.

In all, the second unethical behavior illustrates the extreme importance of the assumptions and premises, which are usually adopted in substantial numbers during the execution of a detailed risk analysis. There is a vast opportunity to manipulate these assumptions and premises in an unethical manner. It is, therefore, extremely important that all risk analysis contractors are aware of this aspect and take all reasonable steps in order to avoid such actions.

It is also considered to be unlikely that a scope of work not in compliance with applicable regulations would be accepted if the risk analysis contractor is well aware of their ethical obligations. It is therefore important that such issues are emphasized in the education of risk analysis specialists and in the training of risk analysis contractors. Awareness is important, but practical guidance on how these principles apply in practice is crucial.

10.5.3 Ethical Challenges for Authorities

The ethical demands on the regulatory authorities are also significant due to their supervision power and obligations. There are two main forms of unethical behavior. On the one hand, the regulatory authorities insist on approving the plan of hazardous facilities even though serious failures and deficiencies on the risk assessment studies have been found and known without independent review. On the other hand, the regulatory authorities conceal the negative conclusion from the verification of the third party when the neighbors' appeal is rejected.

Most importantly, if the interests of some officials and the enterprises overlap, the unethical behavior from regulatory authorities will be hardly avoided. Thus, anti-corruption education and training in the work practice for officials are of great importance in the face of ethical challenges.

10.6 Misuse of Risk Analysis

In practice, several cases have occurred in the offshore petroleum industry in which risk calculations were used as part of the basis for deviating from the specific requirements in the regulations. This practice or interpretation of the regulatory framework is not consistent with the HES principles in the regulations. In this context, Havtil reiterated their warnings with respect to the misuse of risk analysis in November 2017 (Havtil 2017). There are two specific situations that Havtil has focused on:

- Unacceptable use of risk calculations as an argument for deviating from specific requirements in the regulations relating to health, environment and safety.
- Unacceptable use of risk calculations as an argument for choosing solutions that give an inferior safety level in comparison to the established minimum level.

As an example of the first situation, the specific requirement to have fire partitions between main areas on an installation is mentioned in the regulation, and thus this requirement is applicable even if the risk analysis results could demonstrate that the annual frequencies of all fire scenarios are below 10^{-4} per year. Adherence to accepted norms in the offshore petroleum industry is used as an example of the second situation, which possible would prefer solutions that give an inferior safety level.

Havtil also presents the purpose of carrying out risk analysis on the NCS as follows:

- Identifying what are the most important contribution to risk are for personnel, environment and assets.
- Identifying new and improved solutions,
- Analyzing and evaluating the risk level associated with specific technical solutions and activities, both individually and cumulatively.

Further, it is emphasized by Havtil that challenging of accepted solutions is not prohibited, as long as the objective is to improve the safety level further. It becomes misuse of risk analysis only when the safety level is deteriorated with the underlying ambition to reduce the cost.

10.7 Questions

Q1: What is the alternative to the Norwegian trust-based system where enterprises are responsible for safety?

Q2: What are the advantages and disadvantages for the Norwegian system in comparison to an alternative system?

Q3: Should there be a requirement in legislation for risk analysis consultants to be impartial and neutral vis-à-vis all stakeholders?

Q4: Please describe the most important requirements for the ethical behavior of enterprises and risk analysis consultants when major hazard facilities are being planned.

Q5: How can regulatory authorities with a governance system based on internal control principles ensure ethical behavior of all stakeholders when major hazard facilities are being planned?

References

Aven T, Kristensen V (2005) Perspectives on risk: review and discussion of the basis for establishing a unified holistic approach. Reliab Eng Syst Saf 90:1–14

Aven T, Vinnem JE (2007) Risk management with application from the offshore petroleum industry. Springer

Capra G, Cleaver P, Chester A, Phillips A (2007) QRA of the proposed Lyse Gass LNG base load export terminal. Report No: R100-PB-S-SR0001

Danielsen HK (2007) Risavika LNG—third party review of QRA. DNV Memo No HKDA/228703

Havtil (2017) About unacceptable use of risk calculations in order to deviate from requirements in the HES legislation. Letter to oil companies and contractors

Havtil (2020) Regulations relating to management and the duty to provide information in the petroleum activities and at certain onshore facilities. Last amended 16 Dec 2020

Hopkins A, Hale A (2002) Issues in the regulation. In: Kirwan B, Hale A, Hopkins A (eds) Changing regulation: controlling risks in society. Pergamon, Amsterdam

HSE (2008) Guidance on 'as low as reasonably practicable' (ALARP) decisions in control of major accident hazards (COMAH)

Jacobsen SR, Dalsgaard LJ, Etterlid H (2008) Investigation of incident, hydrocarbon leak in utility shaft on Stafjord A 24.05.2008 (in Norwegian). Petroleum Safety Authority Norway

Karlsen JE, Lindøe PH (2006) The Nordic OHS model at a turning point? Policy Pract Health Saf 4(1):17–30

Linde Engineering (2008) Quantitative risk analysis (QRA) Lyse LNG base load plant Train 1. Report No: R100-LE-S-RS0003

Lindøe P, Braut GS (2009) Risk regulation in the Norwegian petroleum industry: robustness and changing methods of operation. In: Bris R, Guedes Soares C, Martorell S (eds) Reliability, risk and safety, theory and applications. CRC Press, Taylor & Francis Group, London

NATO/PFP UNCLASSIFIED-III-1-1 (2003) Risk acceptance criteria. AASTP1-III

Norway Legislation (2002) Act relating to the prevention of fire, explosion and accidents involving hazardous substances and the fire services. Legislation no. 17

Schiefloe PM (2005) Cause analysis for the Snorre A incident. StatoilHydro report, 28 Nov 2004 (in Norwegian)

Standards Norway (2008) Requirements for risk assessment (in Norwegian), NS 5814

Standards Norway (2023) Risk and emergency preparedness assessment. NORSOK Z-013. Rev.4

Uijt de Haag P, Gooijer L, Frijns PJMG (2008) Quantitative risk calculation for land use decisions: the validity and the need for unification. In: The 9th international conference on probabilistic safety assessment and management, Hong Kong, China

Vatn J, Vatn GÅ, Drottz-Sjøber BM (2008) Societal security—a case study related to an LNG facility. Social Security Conference, Norwegian Research Foundation

Vinnem JE (2010a) Risk analysis and risk acceptance criteria in the planning processes of hazardous facilities—a case of an LNG plant in an urban area. Reliab Eng Syst Saf 95:662–670

Vinnem JE (2010b) On the ethics of risk assessment. In: The 20th European safety and reliability conference (ESREL 2010), Rhodos, Greece

Glossary

To facilitate readers' understanding and rapid reference to relevant terminologies in this book, the important terms and definitions are systematically compiled as follows, primarily on the basis of the 2010 and 2023 editions of the NORSOK Z-013 standard, along with relevant technical proposals and reports from Havtil. Notably, these definitions are technically aligned with recognized ISO terminologies, particularly in relation to ISO/IEC Guide 73:2002, ISO 17776:2016, and ISO 31000:2018. Furthermore, certain important terms and corresponding definitions from the Health and Safety Executive (HSE 2001), the International Maritime Organization (IMO 2002), the International Electrotechnical Commission (IEC 2010) and the International Marine Contractors Association (IMCA 2002) are also included for reference.

Accident An unintended event that results in loss of life or damage to health, assets or the environment (adapted from IMO 2002)

Accident scenario A sequence of events from the initiating event to one of the undesirable consequences (adapted from IMO 2002)

Accidental effect The result of an accidental event, expressed as heat flux, impact force or energy, acceleration, etc., which is the basis for the safety evaluation

Accidental event An event or a chain of events that may cause loss of life or damage to health, assets or the environment

Active barrier A barrier that relies on activation or intervention to execute its intended function

X. Zhen and J.-E. Vinnem, *Ten Lessons on Risks*,
https://doi.org/10.1007/978-3-031-99064-9

As low as reasonably practicable (ALARP)	ALARP expresses that the risk shall be reduced to a level that is as low as reasonably practicable
Assumption	By assumptions, we understand conditions/inputs that are fixed in the assessment, but which are acknowledged or known to possibly deviate to a greater or lesser extent in reality
Availability	The ability of an item to be kept in a functioning state. The availability of an item depends on its reliability, maintainability and maintenance support The proportion of time for which an item is working or fit for work (IMCA 2002)
Average individual risk (AIR)	The risk an average individual is exposed to during a defined period of time
Barrier element	Technical, operational and organizational measures or solutions involved in the realization of a barrier function
Barrier function	The task or role of a barrier
Barrier management	Coordinated activities for establishing and maintaining barriers so that they fulfil their functions at all times
Barrier strategy	Documentation that records how specific hazards or accidental events are managed by the use of barrier functions, barrier systems and barrier elements
Barrier system	A system designed and implemented to perform one or more barrier functions
Barrier/safety barrier	A measure intended to identify conditions that may lead to failure, hazard and accident situations, prevent an actual sequence of events from occurring or developing, influence a sequence of events in a deliberate way, or limit damage and/or loss
Basic event	An event at the lowest level of a fault tree
Causal analysis	The process of determining potential combinations of circumstances leading to a top event
Consequence	Outcome of an event affecting objectives. Note that a consequence can have positive or negative effects on objectives
Context	Factors in the surroundings that are relevant for assessing safety and the need for barriers

Control	Measure that maintains and/or modifies risk
Cost–benefit analysis	Quantitative analysis and comparison of costs and benefits
Defined situations of hazard and accident (DSHA)	Selection of hazardous and accidental events that will be used for the dimensioning of the emergency preparedness for the activity
Design accidental load	Accidental load/action used as a basis for design
Dimensioning accidental event	Accidental events that serve as the basis for layout, dimensioning and use of installations and the activity at large
Dimensioning accidental load	Accidental load/action that a function or a system shall be able to withstand for a given period of time to meet the defined acceptance criteria for risk and minimum requirements in the regulations
Emergency preparedness	Preparations to respond to critical conditions. Specifically, it refers to technical, operational and organizational measures, including necessary equipment that are planned to be used under the management of the emergency organization in case hazardous or accidental situations occur, in order to protect human and environmental resources and assets
Emergency preparedness analysis	Analysis which includes identification and establishment of analysis basis, establishment of DSHAs, establishment of emergency response strategies, establishment of emergency performance requirements, dimensioning the emergency response organization, documentation of process and results, including recommendations. Note that emergency preparedness assessment does not include establishment of emergency preparedness
Emergency preparedness assessment	The overall process of performing an emergency preparedness assessment including: establishment of the context, emergency preparedness analysis, emergency response plan, training and exercise, follow-up and continuous improvement. Establishment of emergency preparedness represents a systematic process that involves selection and planning of suitable

	emergency preparedness measures on the basis of risk and emergency preparedness analysis
Emergency preparedness philosophy	Overall guidelines and principles for the establishment of emergency response based on the operator vision, goals, values and principles
Emergency response	Action taken by personnel on or off the installation to control or mitigate a hazardous event or initiate and execute abandonment
Emergency response strategy	A specific description of emergency response actions for each DSHA. Note that emergency response strategies shall be the basis for the establishment of the emergency response plan
Environment	Surroundings in which an organization operates, including air, water, land, natural resources, flora, fauna, humans and their interrelation
Environmental impact	Any change to the environment, whether adverse or beneficial, wholly or partially resulting from an organization's activities, products or services
Escalation	Fire or explosion or other hazardous events that affect other fire area(s) than that of the initial event, i.e., either the integrity of one or more fire divisions between fire areas is impaired (do not fulfil the intended function), or hazardous substances in the other fire area(s) are released
Escape	Actions by personnel on board surface installations (as well as those by divers) taken to avoid the area of accident origin and accident consequences to reach an area where they may remain in shelter
Escape route	Route from an area of an installation leading to a muster or embarkation area
Emergency response organization	An organization which is planned, established and trained in order to handle occurrences of hazardous or accidental situations
Evacuation	Planned method of leaving the facility in an emergency situation
Event	Occurrence or change of a particular set of circumstances

Event tree analysis Inductive analysis in order to determine alternative potential scenarios arising from a particular hazardous event. It may be used quantitatively to determine the probability or frequency of different consequences arising from the hazardous event

Explosion load Time-dependent pressure or drag forces generated by the violent combustion of a flammable atmosphere

Facility Offshore or onshore petroleum installation, facility or plant for the production of oil and gas

Failure Inability of an item to execute its required function, either temporarily or permanently

Failure mode The way in which an item could fail to perform its desired function (IMCA 2002). Sometimes, it is described as categories of failures

Fault tree analysis A deductive quantitative analysis technique in order to identify the causes of failures and accidents and quantify the probability of their occurrence

Fire area Area separated from other areas either by physical barriers (fire/blast division) or sufficient distance, which will prevent a fire or explosion to escalate to another fire area when exposed to the design loads

Frequency The number of occurrences per unit time

Functionality of barriers The influence that the barriers have on the chain of events (assuming their presence)

Grossly disproportionate The bias in favor of safety when assessing that is reasonably practicable (HSE 2001)

Harm Injury or damage to the health of people, or damage to property or the environment

Hazard Potential source of harm. Note that the potential harm may relate to loss of life, or damage to health, the environment or assets or a combination of these

Hazardous event An incident which occurs when a hazard is realized

Individual risk The risk an individual is exposed to during a defined period of time. Note that the individual risk may be established by combining the fraction of time an individual is located in various areas and the area risk in each of the areas

Inherent safe design — Philosophy that focuses on the elimination of hazards or the reduction of the magnitude of hazards rather than the control of hazards. The following concepts are used to reduce risk:

- reduction, e.g., reducing the hazardous inventories or the frequency or duration of exposure
- substitution, e.g., substituting hazardous materials with less hazardous ones (but recognizing that there could be some trade-offs here between plant safety and the wider product and lifecycle issues)
- moderation, e.g., using the hazardous materials or processes in a way that limits their hazard potential, such as segregating the process plant into smaller sections using ESD valves, processing at lower temperature or pressure
- simplification, e.g., making the plant and process simpler to design, build and operate, hence less prone to equipment, control and human failure

Initiating event — The first event of a specific sequence of events resulting in a hazardous situation or accident (adapted from IMO 2002)

Integrity — The ability to be in place and intact at all times

Internal control — All administrative measures which are implemented to ensure that the work is in accordance with all requirements and specifications

Likelihood — Chance of something happening

Main area — Facility area intended for a specific task or function

Main load-bearing structures — Structure, which when it loses its main load-carrying capacity, may result in a collapse or loss of either the main structure of the installation or the main support frames for the deck

Main safety function — The most important safety functions that need to be intact in order to ensure the safety for personnel and/or to limit pollution

Maintainability	The ease with which a failed item may be repaired (IMCA 2002). The usual measures are the mean time or the distribution of time to repair
Maintenance	All actions necessary for retaining an item in, or restoring it to, a serviceable condition. Includes servicing, repair, modification, upgrading, overhaul, inspection and condition determination (IMCA 2002)
Major accident	Acute incident, such as a major discharge, a fire or an explosion, which immediately or later causes several serious personal injuries and/or loss of life and/or serious harm to the environment and/or loss of major material assets
Major hazard	Hazard with the potential, if realized, to result in a major accident
Mitigation	The limitation of any negative consequence of a particular event
Normalization	The normalization phase starts when the development of a situation of hazard or accident has stopped
Occupational accidents	Accidents relating to hazards that are associated with the workplace, including falls, slips, and crushing injuries, rather than hazards involving hydrocarbon gas or oil under pressure. These accidents are normally related to a single individual
Operational barrier element	The actions or activities which personnel must perform in order to realize a barrier function
Organizational barrier element	Personnel with defined roles or functions and specific competence involved in the realization of a barrier function
Passive barrier	A barrier that does not rely on activation or intervention to execute its intended function
Performance-influencing factors for barriers	Factors identified as having significance for barrier functions and the ability of barrier elements to function as intended
Performance requirements for barriers	Verifiable requirement for the properties of the barrier elements in order to ensure that the barrier is effective

Performance standard	Measurable statement, expressed in qualitative or quantitative terms, of the performance required of a system, item of equipment, person or procedure, and that is relied upon as a basis for managing a hazard
Personnel safety	Safety for all personnel involved in the operation of a field
Precursor events/Near accident/Near miss	An event that under other circumstances could reasonably have developed into major accidents, but actually did not
Probability	The extent to which an event is likely to occur
Quality	A concept which embodies variously, and as appropriate, the ideas of performance (or fitness for purpose), durability, freedom from repairable failure, maintainability, and even aesthetics (IMCA 2002). It does not include any consideration of price or cost
Reasonably practicable	The costs (in terms of money, time, etc.) involved in implementing a measure are not grossly disproportionate to the benefits obtained (HSE 2001)
Recovery time	Time from an accidental event causing environmental damage occurs until the biological features have recovered to a pre-spill state or to a new stable state, taking into consideration natural ecological variations, and are providing ecosystem services comparable to the pre-spill services
Reliability	The ability of an item to perform a specific function (under given conditions) for a stated period of time (IMCA 2002)
Residual risk	Risk remaining after risk treatment
Risk	Consequences of the activities, with associated uncertainty. In tradition, risk represents the combination of the probability of occurrence of harm and the severity of that harm
Risk acceptance	Decision to accept risk
Risk analysis	Structured use of available information to identify hazards and to describe risk. Note that the risk analysis term covers several types of analyses that will all assess causes for and consequences of accidental events, with respect to risk to personnel, environment and assets. Examples of the simpler analyses include

	FMEA, PHA, HAZOP, among others. Quantitative analysis may be the most relevant in many cases, involving a quantification of the probability and the consequences of accidental events, in a manner which allows comparison with quantitative RTC
Risk assessment	The overall process of establishment of the context, risk analysis, risk evaluation
Risk awareness	The recognition and understanding of potential risks
Risk control	Actions implementing risk management decisions
Risk criteria	Terms of reference against which the significance of risk is evaluated
Risk identification	Process to seek for, list and characterize elements of risk
Risk evaluation	Judgement, on the basis of risk analysis and RTC, of whether a risk is tolerable or not
Risk indicator	A measurement or value that is developed from risk analysis models
Risk-influencing factor	A relatively stable factor that influences the risk
Risk management	Coordinated activities to direct and control an organization with regard to risk
Risk perception	The subjective judgment with respect to the characteristics of risk, based on a set of values or concerns
Risk picture	Synthesis of the risk assessment, with the intention to provide useful and understandable information to relevant decision makers
Risk reduction	Actions taken to reduce the probability, negative consequences, or both, associated with a hazard
Risk tolerance	Organization's readiness to bear the risk after risk treatment in order to achieve its objectives
Risk tolerance criteria (RTC)	Criteria that are used to express a risk level that is considered the upper limit for the activity in question to be tolerable. Note that risk tolerance criteria are the starting point for further risk reduction according to the ALARP principle
Risk transfer	Sharing with another party the potential burden of loss or benefit of gain, such as contracts and insurance

Risk treatment Process of selection and implementation of measures to modify risk

Robustness The ability of a system and an organization to maintain their function when circumstances change and in the event of failures, hazards and accidents. Note that robustness is the opposite of vulnerability, which deals with the inability to handle such circumstances

Rooms of significance Rooms of significance to combating accidental events

Root cause The most basic cause that can be reasonably identified, and that management has control over to fix. If root causes are addressed, recurrence of similar failures can be prevented

Safe area(s) Area(s) that remain intact until the facility has been evacuated or accident is normalized

Safety

(a) From a risk science perspective, it is a state where risk has been reduced to a level that is ALARP and the remaining risk is generally tolerable
(b) freedom from unacceptable risk (IEC 2010)

Safety case A comprehensive document providing a compelling, comprehensible and valid case of how the duty holder intends to manage a system or an installation safely. The idea of a safety case is to provide an argument similar to what one would do in a court of law

Safety culture The product of individual and group values, attitudes, competencies, and patterns of behavior that determine commitment to, and the style and proficiency of, an organization's health and safety management. Organizations with a positive safety culture are characterized by communications founded on mutual trust, by shared perceptions of the importance of safety, and by confidence in the efficacy of preventive measures (HSE 2005)

Safety function Function to be implemented by a safety-related system or other risk-reducing measures, that is intended to achieve or maintain a safe state, in respect of a specific hazardous event (IEC 2010)

Safety integrity	Probability of a safety-related system satisfactorily performing the specified safety functions under all the stated conditions within a stated period of time. Notably, there are four levels of safety integrity, and the higher the level of integrity, the lower the probability that the safety-related system will fail to carry out the specified functions (IEC 2010)
Safety integrity level	Discrete level (one out of a possible four), corresponding to a range of safety integrity values, where safety integrity level 4 has the highest level of safety integrity and safety integrity level 1 has the lowest (IEC 2010)
Scenario	A sequence of events
Shelter area	An area on the platform where the crew will remain safe for a specific period of time in an emergency situation
Stakeholder	Person or organization that can affect, be affected by, or perceive themselves to be affected by a decision or activity
Structural integrity	Load carrying capacity in main load-bearing structures
System	Common expression for installation(s), plant(s), system(s), activity/activities, operation(s) and/or phase(s) subjected to the risk and/or emergency preparedness assessment
System boundaries	System boundaries define what shall and what shall not be subjected to the assessment
Technical barrier element	Equipment and systems involved in the realization of a barrier function
Uncertainty	Not knowing the true value of a quantity or the future consequences of an activity. Imperfect or incomplete information/knowledge about a hypothesis, a quantity, or the occurrence of an event. Note that it is common to distinguish aleatory uncertainty (irreducible uncertainty stemming from real variation of quantities) and epistemic uncertainty (uncertainty that may be reduced through acquisition of knowledge)
Vulnerability	The lacking ability of an object to resist the impacts of an unwanted event and to restore to its original state or function following the event (Rausand 2011)

References

HSE (2001) Marine risk assessment. HMSO, London

HSE (2oo5) A review of safety culture and safety climate literature for the development of the safety culture inspection toolkit. Report No: 367, Health and Safety Executive, London

IEC (2010) Functional safety of electrical/electronic/programmable electronic safety-related systems-part 4: definitions and abbreviations. IEC 61508, Edition 2.0, Geneva

IMCA (2002) Guidance on failure modes & effects analyses (FMEAs). Report No: IMCA-M-166

IMO (2002) Guide for formal safety assessment (FSA) for use in the IMO rule-making process. Report No: MSC/1023. International Maritime Organization, London

ISO (2009) Risk management—vocabulary. ISO/IEC guide 73. International Standards Organization, Geneva

ISO (2016) Offshore production installations—major accident hazard management during the design of new installations. ISO 17776. International Standards Organization, Geneva

ISO (2018) Risk management—guidelines. ISO 31000. International Standards Organization, Geneva

Rausand M (2011) Risk assessment: theory, methods, and applications. Wiley, Hoboken, New Jersey

Standards Norway (2010) Risk and emergency preparedness assessment. NORSOK Z-013. Rev.3

Standards Norway (2023) Risk and emergency preparedness assessment. NORSOK Z-013. Rev.4

Index

X. Zhen and J.-E. Vinnem, *Ten Lessons on Risks*,
https://doi.org/10.1007/978-3-031-99064-9

R

S

T

MIX
Papier aus verantwortungsvollen Quellen
Paper from responsible sources
FSC® C105338

If you have any concerns about our products,
you can contact us on
ProductSafety@springernature.com

In case Publisher is established outside the EU,
the EU authorized representative is:
Springer Nature Customer Service Center GmbH
Europaplatz 3, 69115 Heidelberg, Germany

Printed by Libri Plureos GmbH
in Hamburg, Germany